Principles of Water Management in Agriculture

Principles of Water Management in Agriculture

Klein Gomes

Oxford Book Company
Jaipur ▪ New Delhi

Published By

Oxford Book Company

Regd. Off. "Bony Residency", Gate No. 2
Opp. Tilak Public School, Vishweswariya Nagar,
Gopalpura Road, Jaipur - 302018, India
Ph: 0141 - 2761280, Telefax: 2761381
email - oxfordbookjpr@gmail.com
website - oxfordbookcompany.com

Delhi Office: 102, 1st Floor, Satyam House,
4327/3, Ansari Road, Darya Ganj,
New Delhi - 110002, India
Ph: 011-45652440

Published in India, 2025

ISBN : 9789380179018

Price : ₹ 2250/-

Typeset by Shivangi Computers, Jaipur

Printed and bound in India by Thomson Press Ltd.,
New Delhi

Preface

A worldwide attention is sharply focused, nowadays, on seeking ways of achieving the efficient and equitable use of water resources. The food security concerns obviously stem from looming threat of water scarcity. Effective management of water for agriculture, links socio-economic development with the protection of natural ecosystems. The absorbing topics on agriculture and environmental threat of water scarcity, techniques of water management of field crops, remote sensing in crop water management, soil and water conservation, ecological vulnerability of wetland ecosystems, adoption of micro-irrigation system for efficient water management and increasing farm productivity, water resources management and climate, phenomenon of muddy precipitation and pollution of water bodies are various facts of water management for agriculture.

Technologies for water management provide a paradigm shift ensuring supply of water to every field in the rainfed areas and this would remove hunger and poverty from the devastating dry land/drought-prone areas, restoring ecological balance, providing green cover over denuded areas, bringing in more rains, and improving the environmental, economy, ecology, export and equity in the country. Uniformity is not compatible with land, crop husbandry, and diversified mixed farming systems for each plot of land and each household would necessarily need different treatments and dynamic combination in consonance with natural endorsement and socio-economic needs.

In the modern world, the use of gene manipulation, better management of soil nutrients, and improved weed control have greatly increased yields per unit area. Knowledge, both traditional and new, has been critical to meet the food needs of growing populations. Innovations in agriculture may also help address long-standing issues such as land degradation, and pests and disease control, as well as new challenges such as climate change and water scarcity. The success of agricultural development will depend largely on the support and participation of rural people, national governments, the private sector and international cooperation, including technical and scientific cooperation.

Klein Gomes

Contents

Chapter 1

The Global Challenge

In his *History of the World* Roberts, commented on a world population exceeding 5 billion persons, "Though it had taken at least 50,000 years for *Homo sapiens* to increase to 10,00 millions, the last 1,000 millions of his species took only 15 years to be added to a total growing more and more rapidly". More recently it is estimated that the world population reached 6 billion in October 1999 and that the increase from 5 to 6 billion, huge army of the world's desires, took place in a mere 12 years. More than any other single factor, it is these prodigious strides in world population size that in the early years of the third millennium are the source of anxiety about the future of the world's water resources available *to Homo sapiens.*

The reason is clear: whereas world population will increase by about one-third to some 8 billion by the year 2025, there is no prospect whatsoever of *any* increase in the global effective rainfall that nurtures the world's plant and animal life and that feeds its lakes, rivers and aquifers. It is David Seckler and his colleagues at the International Water Management Institute in Sri Lanka who have contributed most to our understanding of the changing relationship between the global water supply and its demand that can be expected by 2025. Using their policy interactive dialogue model they present a "basic scenario" that is shaped by their four major objectives, which aim:

- To achieve an adequate level of per capita food consumption, partly through increased irrigation, to reduce substantially malnutrition and the most extreme forms of poverty.
- To provide sufficient water to the domestic and industrial sectors to meet basic needs and economic demands for water in 2025.
- To increase food security and rural income in countries where a large percentage of poor people depend on agriculture for their livelihoods through agricultural development and protection from excessive and often highly subsidized agricultural imports.
- To introduce and enforce strong policies and programmes to increase water quality and support environmental uses of water.

The data available to IWMI were sufficient in the case of the forty-five countries that encompass 83 per cent of the world's 1995 population that the PODIUM model–with all its inevitable predictive uncertainties–could be used to allocate each of these nations to one of three groups.

Group I is composed of countries likely by 2025 not to have sufficient water resources to meet their agricultural, domestic, industrial and environmental needs. It contains Algeria, Egypt, Iran, Iraq, Israel, Jordan, Pakistan, Saudi Arabia, South Africa, Syria, Tunisia, plus that one-half of the population of China living in the arid north, mainly in the Yellow River basin and that one-half of the population of India living in the arid north-west and south-east.

Group II countries are predicted to have to increase their water supplies through additional storage, conveyance and regulation systems by at least 25 per cent over 1995 levels to meet their 2025 needs. These are Argentina, Australia, Bangladesh, Brazil, Ethiopia, Mexico, Morocco, Myanmar, Nigeria, the Philippines, Sudan, Thailand, Turkey and Vietnam plus that one-half of the populations of China and India not included in group I.

Group III countries are predicted to need to increase their water supplies by less than 25 per cent over 1995 levels to meet their 2025 needs. They are Canada, France, Germany, Indonesia, Italy, Japan, Kazakhstan, Kyrgyzstan, Poland, Romania, the Russian Federation, Spain, Tajikistan, Turkmenistan, the Ukraine, the UK, the USA and Uzbekistan.

Groups I and II, the "physical water scarcity" and "economic water scarcity" countries, will make up, respectively, 33 per cent and 45 per cent of the 2025 total population of the forty-five countries categorized. The estimated total global population used is in the "basic scenario".

In addition to the increase in the world's population, in the early years of the new millennium there is a second source of anxiety about the future availability of global water resources to humankind. This concerns climate change, for which the predictive time-scale is much longer than that used for population growth.

The limiting case is taken where emissions of CO_2 into the atmosphere are not mitigated between the years 2000 and 2080 with consequential effects on global warming. Even mitigating action needs decades to take significant effect.

The approach simulates river run-off across the globe at a spatial resolution of 0.5° × 0.5° latitude and longitude. The macroscale hydrological model was run firstly with the observed 1961-90 global climate, followed by scenarios derived from predictions made by the Hadley Centre in the UK. But the projections take no account of the effect of climate change on the *demand* for water.

Arnell points out that, at the turn of the millennium, 1.7 billion people lived in countries that were experiencing water stress, i.e., where more than 20 per cent of the average annual renewable resource is used. Climate change has the potential to alter these patterns of stress. Some parts of the world will experience an increase in river run-off that may be accompanied by increased flooding. Important cases are North America, Asia and central eastern Africa. But substantial decreases are seen in Australia, India, southern Africa, most of South America and Europe and the Middle East.

On the basis of this work, it appears possible that, in the life-time of our children and grandchildren, countries such as the USA, China and Uganda will see improved conditions for agriculture as well as the greater availability of irrigation water. The reverse is true, for example, in India, South Africa and Egypt. Another perspective on global warming is given in a visionary document on water for food and rural development financed by the Netherlands Directorate General for Development Cooperation. Here the orientation is specific to farming.

The gradual warming of the earth, 1.0°C in the past 50 years, is leading to glacial recession, declining snow cover and rising sea levels. Precipitation patterns are likely to alter, reducing water availability in some regions and increasing it in others. Increased variability in precipitation patterns will accompany this shift with a huge impact on both irrigated and non-irrigated agriculture.

Precipitation patterns will include a greater proportion of extreme events, leading to higher and more frequent flooding and lower dry season flows in rivers. More intense rainfall will lead to increased erosion and higher sedimentation rates in reservoirs and canals. The production potential of past investments in water control facilities will be reduced where reservoirs can no longer be filled due to decreased precipitation.

THE IRRIGATION CYCLE

Up to this point in the text, references to the principal activity with which this book is concerned-agricultural irrigation–have been sparse. The moment has now come to start to address this subject. In fact the PODIUM model has a particularly strong orientation towards farming, for in the year 2000 the total cultivated area of the world was 1 billion hectares, of which more than one-quarter was irrigated outlines some of the salient information.

The last column shows for the year 2025 just what a great proportion of the world's water resources diverted for human use is likely to be channelled into agriculture. Irrigation as a percentage of total diversions ranges from as little as 2-4 per cent in Canada, Germany, Poland and the UK up to 90-95 per cent in Iraq, Pakistan, Bangladesh, the Sudan,

Kyrgyzstan and Turkmenistan, with an overall average of 68 per cent. In terms of predicted 2025 irrigation diversions the five big fish in order of volume swallowed are India, China, Pakistan, the USA and Mexico. They take 69 per cent of the 2,786 km^3 diverted. Next comes a group of twenty-six countries with an off-take of between 10 and 85 km^3 and finally the fourteen minnows with diversions of less than 10 km^3.

Hydroeconomics demonstrates how strong in quantitative terms is the argument for a specific economics of irrigation. But what should be the content of such a subject?

The brief answer is that the economics of irrigation is the application of economic theory to *the irrigation cycle.* A simple model of the irrigation cycle refers to a dynamic somewhere along the continuum between the hydrological processes of the natural world, such as precipitation and the hydrosocial activities of humankind, such as pumping drainage water to a field.

This fundamental synegy of hydrological and hydrosocial forces, this conjunction and opposition of two spheres, exists even if the relationship between the two process types is complexly interdependent and, for that reason, one is not always clearly demarcated from the other. The starting point is the hydrological resource at the catchment scale composed of rainfall, freshwater lakes, rivers and aquifers. This resource is drawn upon by four processes:

- Evaporation from lake and river.
- Run-off to the oceans and to land-locked saline sinks.
- Abstraction for household, industrial and other such uses.
- Appropriation for irrigated agriculture. The flows to irrigation are made up of five categories that are referred to throughout this book as *the five base flows:*
  - Rainwater collection prior to the run-off phase of the hydrological cycle. This takes place in a variety of small-scale forms including its use for microscale kitchen gardens.
  - Water abstracted from rivers and lakes. This may be by diversion of the surface-water flow such as in a river's upper reaches or from downstream barrages permitting large areas to be commanded by irrigation canals, or by lift pumps from river and lake. Here are also included phenomena such as the natural summer rise of the Nile in Egypt that, with its flat riparian areas, is the basis of a simple form of flood irrigation that has been practised for thousands of years.
  - Water abstracted from aquifers, including the capture of spring flows.
  - The reuse of household and urban wastewater.

- The reuse of irrigation water itself, taken from drainage channels.

Desalination as a source is considered here to be of such negligible importance that it is not included. That may change in the future.

At the global level surface-water and groundwater abstraction are by far the most important categories, but they may not be so at the local level.

Note that these two base flows may themselves be subdivided into virgin and recycled water. "Recycled water" is water that, within the limits of a given year, has already been abstracted and then is returned to the hydrological resource as agricultural drainage or as the wastewater of households and industry. "Virgin water" is water abstracted from the catchment resource that is not recycled water. "Reused water", categories 4 and 5 of the base flows above, is wastewater and agricultural drainage water that is captured and applied for irrigation prior to its return to the hydrological resource.

This important general distinction between recycling and reuse is used repeatedly in this text. Recycling refers to water that is appropriated during the course of the year as one of the base flows and then returns, by one pathway or another, to the hydrological resource, i.e. the catchment's rivers, aquifers and lakes.

The importance of recycling to hydroeconomists is that it augments the hydrological resource from the point at which the recycling occurs. The negative characteristic is that recycled flows may pollute the resource. Abstraction downstream of the recycling point will be composed in part of recycled water and in part of virgin water. This subject has been dealt with at length in the work of Perry.

Reuse, in contrast, refers to primary flows that after appropriation are first used in the domestic, industrial and irrigation sectors and then are used for a second time or, indeed, time and time again. *Internal* reuse occurs when the repeated use is by the same institution where the water was first used.

This is familiar in the industrial sector and in the domestic sector where grey water is reused to flush toilets. *External* reuse takes place when the repeated use is by a different institution. The importance of internal and external reuse to the hydroeconomist is that it multiplies the productivity of a given volume of water appropriated as a base flow. As with recycling, water quality is a critical issue, in this case the quality of the effluent reused.

The five base flows to irrigation usually need to be channelled to and stored in small- or large-scale reservoirs so that their availability to farmers can be secured at the right time of the changing seasons. Aquifers themselves may be used as natural reservoirs and "recharged" purposely for later use. When the time is right, the contents of the reservoirs are distributed to the fields for application to the growing crops. The supply of water over the life of a dam can be expected to decline as siltation removes storage capacity. A heavy silt load affects the economics of dams by causing abrasion to structures and it also increases the operating costs of canals.

But as already indicated, the base flows are not used up entirely in agriculture as crop evapotranspiration. Between the starting point of each flow and the point of crop take-up, a series of "losses" occur. These are: evaporation such as from dams, canals and open fields; non-crop transpiration; leaching applications to cleanse the soil of accumulated salts; escapes; seepage from storage and the distribution network; deep percolation during and after application of the irrigation water at the field level; and surface run-off. All but the first two of these "losses" return as drainage to the hydrological resource, whence they may be recycled to agriculture and other human uses.

The most direct example of this cyclical movement is where seepage and deep percolation recharge the aquifer that supplies the same irrigation district whence the original losses took place. In many catchments "losses" can be categorized as those: (1) between the point at which the base flow is appropriated and the irrigation district boundary; (2) between that boundary and the access point to the farmholding; (3) between the access point and the irrigated field; and (4) "losses" at the field level itself. The term "losses" is put in quotation marks here because the only loss to the hydrological resource at the catchment level is through evapotranspiration–and even some part of this may return through precipitation, particularly in large-scale catchments such as the Amazon, the Congo, the Ganges and the Mekong.

So the economics of irrigation can be defined as the application of political economy to: the *appropriation* of water for agriculture as five base flows; irrigation water *storage* and *distribution* to and from storage of these flows; their *use* in agriculture; and *drainage* back to the hydrological resource of that part of the flow not consumed by evapotranspiration. Each flow and sub-flow of the irrigation cycle can be described quantitatively in terms of its volume per unit time and qualitatively in terms of its organic, inorganic, microbiological and other characteristics. A final point needs to be made here in respect of the term "cycle". The attraction of concepts such as hydrological cycle, hydrosocial cycle and irrigation cycle is that they give strong emphasis to the return flows of recycling and reuse within a catchment. However, it should be remembered that such cycles are only *partial*. Even in the paradigm case of the hydrological cycle, the catchment exports and imports water to and from other catchments via the processes of run-off to the oceans, evaporation and precipitation. It is only at the global level that the hydrological cycle is complete rather than partial. With respect to the irrigation cycle, this deficit characteristic is of particular importance in two cases:

- Where the abstraction rate from an aquifer in fuelling irrigation supply is so great that the aquifer is progressively exhausted.
- Where the base flows required for the irrigation sector are so great that they can be met only by the import of water from another catchment.

These are issues to which we shall return repeatedly in this book's empirical content.

THE ECONOMIC PARADIGM

If water is so fundamental a biological requirement in agriculture, if irrigation water (and other outstream flows) is now widely recognized to be an economic good and if irrigation water constitutes about 70 per cent of all diversions, then there is a need for an economics of irrigation. There is also a potentially varied audience with an interest in that subject. This book is intended to meet the needs of both students and professionals in the fields of agriculture, development studies, economics, engineering, environmental science, environmental studies, geography, hydrology and planning.

As an introduction to the subject, no prior knowledge of economic theory is assumed. The economic paradigm that provides the analytical foundation of this book deserves mention. This is institutional economics, rather than Marxist economics or neoclassical economics-the two dominant paradigms of the last century. But my approach is open-minded and eclectic: I borrow from these last two paradigms whenever they seem to open a door to understanding. Important expositional texts on institutional economics are Hodgson, Earl and Stretton. For convenience, institutional economics can be defined as the study of human agency and institutions so as to illuminate the concrete processes of social life with respect to the production and allocation of goods and services. Its analysis is open-ended and interdisciplinary in that it draws upon relevant material in psychology, sociology, anthropology, politics and history, as well as economics itself. The conception of the economy is of a cumulative process unfolding in historical time in which agents are faced with chronic information problems and radical uncertainty about the future.

The concern is to address and encompass the interactive, social process through which tastes are formed and changed, the forces which promote technological transformation and the interaction of these elements within the economic system as a whole. It is appropriate to regard the market itself as a social institution, necessarily supported by a network of other social institutions such as the state and having no unqualified or automatic priority over them.

It is recognized that the socio-economic system depends upon and is embedded in, an often-fragile natural environment and a complex ecological system. The enquiry is value-driven and policy-oriented and recognizes the centrality of participatory democratic processes to the identification and evaluation of real needs.

With respect to epistemological approach, I believe that statements about human society should be understood as ranged along a single spectrum from theory at one end to description at the other. Theory is characterized by its abstractness and generality; description by its concreteness and specificity. But theory always informs description and description provides the grounding to theory. So, to succeed, the method of institutional economics should be to maintain a continuing dialectic, a cyclical interdependence, between reflection and fieldwork.

I think of it as a theoempiric procedure, if that is an acceptable neologism. The basis of this separation is that politics, psychology, social theory, etc. are concerned to study humanity, the actions of which are those of reflexive agents who make their own history. This is not true of astronomy, biochemistry, physics, etc. It follows that, whilst the physicist legitimately seeks to identify universal laws, the institutional economist must be content with a more modest interest in generalizations, contingent on time-space and the human heart.

AN IRRIGATION CYCLE IN SOUTHERN INDIA

The second highest population in the world, about three-quarters of whom lived in rural areas and it had the earth's largest irrigated area, in excess of 50 million hectares. So it is particularly appropriate for the first case study of this book to be placed in India. The material is derived from Anna Blomqvist's splendid *Food and Fashion:*

Water Management and Collective Action among Irrigation Farmers

Her fieldwork was carried out over the 5-year period 1992-96. Blomqvist's study was located in the Coimbatore region of western Tamil Nadu in southern India. The main components of the hydrological resource are rainfall, the Bhavani reservoir and the Bhavani River through to its junction with the Cauvery River. The river originates in the Western Ghats and the river-reservoir system provides water to the 16,000 ha of the three Old Ayacuts as well as to the 83,000 ha of irrigated farmland of the Lower Bhavani Project (LBP) first launched in 1956. By the mid-1990s the latter had about 70,000 farmers, each owning on average 1.2 ha. Except for the canal-irrigated areas the region is dry, surrounded as it is by mountains to the north, west and south and affected by the rainshadow they create.

The average annual rainfall of the region is 650 millimetres (mm), distributed between the south-west and north-east monsoons. Annual variation in precipitation is substantial and potential evapotranspiration ranges between 1,500 and 2,000 mm. Approximately 90 per cent of the freshwater resource in 1996 was utilized in the agricultural sector. It shows that, from the irrigator's point of view, the wasted outflows from the hydrological resource of the reservoir and the river downstream of it are evaporation, abstraction for urban use and run-off to the River Cauvery. The main form of appropriation of the resource for irrigation is by diversion at the Bhavani reservoir for the Lower Bhavani Main Canal and diversion from the river for the Arakkan Kottai, Thadapalli and Kalingarayan canals of the "Old Ayacuts". Abstraction from the river using electric- and diesel-motor-driven pumps also takes place.

The split between virgin and recycled sources was unknown, but by 1996 the effluents from a large synthetic textiles mill had been polluting the Bhavani River for several years and water pollution was a high priority for the LBP Agriculturist Association. In spite of repeated demands from farmers that environmental legislation be more strictly enforced, little had been achieved. Both farmers and environmental groups blamed the Pollution Control Boards for slack enforcement of existing legislation. The administrative division of the river basin between Coimbatore and Periyar weakened the farmers' countervailing power. Pollution mainly affects farmers in the Periyar district, but many of the polluting units were in the upstream and more distant, Coimbatore district.

The distribution system begins with the four principal canals. Off these appear branch canals of varying size that take the water to the sluice points. Each sluice serves an area of 20-50 ha along field channels. Gravity is the energy source. During the 8 months of the year when water is diverted to the system, the discharge in the mid-1990s was something less than 5.5 million cubic metres (mcm) per day. This was divided almost equally between the Lower Bhavani Project and the Old Ayacuts, in spite of the fact that the LBP was more than five times the combined size of its three older rivals.

Farmers in the old schemes claimed to be in possession of riparian rights and to have an indispensable priority in water allocation between the two schemes. In the Old Ayacuts the embryos existed in the late 1940s of what was to become the Arakan Kottai Farmers' Welfare Association and a similar Thadapalli institution. They were created to secure riparian rights at the time of the construction of the Bhavani dam. The Kalingarayan Farmers' Development Association was formed as a result of tail-end problems in the canal in 1991. "The main activities of all the organizations in the Old Ayacuts is to act politically to protect their water supply. When water supply is adequate, the organizations remain inactive, but are vitalized during periods of scarcity".

Not unexpectedly the farmers of the LBP Agriculturist Association said water should be re-allocated from the Old Ayacuts to them, particularly as harvests were considerably higher in the LBP. The Old Ayacut response was that their soil would be destroyed if it were to dry out, so water had to be supplied on a 10-month basis.

The Public Works Department (PWD) was responsible for supplying the designed volumes of water, equitably and on time, to the sluice point. But during dry years it was not unusual for insufficient volumes to be supplied even at the main canal level. The PWD should also have maintained and desilted the irrigation infrastructure. At district level, the Superintending Engineer was the highest PWD official with four lower levels of staff below him, all responsible for different components of the infrastructure. The field-level staff, the *lascars,* were in charge of water distribution between sluices in the same branch canal.

Farmers were responsible for the operation, maintenance and distribution of the water below a sluice outlet. The poor reliability of water supply and meagre information about when it would be released created a tendency among farmers to compete for water, leading to over-irrigation at the head-reach and water scarcity at the tail-end. The possibility for individuals or groups of farmers to secure their own water supply by paying bribes caused problems for others.

But these bribes were an important source of income to the underpaid, low-level officials, who encouraged corruption on a regular basis. As one interviewee expressed it: Farmers used to collect money before the season just for paying the lascar. It was a custom. He used some tactics to invite the farmers. If he lowers the shutters, they will approach the lascar and they will have to pay something. The whole system is like that. From the lascar to the Superintending Engineer to the Chief Engineer. They have a target. The use of irrigation water will now be addressed. Originally the LBP had been designed for irrigated dry crops such as cotton, gingelly and groundnuts. Cotton especially was in high demand by the textile industry in the Coimbatore district. The water set aside for the LBP was not enough to grow water-intensive crops such as paddy, sugarcane and turmeric, whereas on the Old Ayacuts farmers were free to cultivate any crop. Soon after the Bhavani reservoir had been completed and the planned area for the new project was taken under irrigation, problems of seepage and illegal water acquisition by head-reach farmers made the enforcement of crop restrictions difficult.

After strong pressure from the farming community, a new principle for irrigation was introduced in 1959. The irrigated area of the LBP was divided into two zones which would be irrigated by either the wet or dry turn every year. Wet crops could be cultivated from mid-August to mid-December. But from mid-December to mid-April, only irrigated dry crops could be grown. These were the plans, but they were not always followed. At the end of the 1980s the gap area in the LBP, i.e., the difference between the potentially irrigated area and the area actually irrigated, varied between 5 per cent and 30 per cent depending on rainfall.

By the mid-1990s irrigated farming was far more important than rain-fed farming. In the Coimbatore district, coconut, sugarcane and groundnuts were the three most important crops, constituting over half of the total output value. In the Periyar district, paddy, sugarcane and turmeric accounted for more than 70 per cent of crop value. In both districts paddy and *jowar* (a type of sorghum) were cultivated on about one-third of the total farmland.

In both the Old Ayacuts and the Lower Bhavani Project there was room for improved water productivity, but most farmers saw reasons for improvement only in the other area, not their own. Anna Blomqvist felt that water productivity, so low compared with industry and the domestic sector, could be raised by reductions of seepage, illegal water acquisition and violation of cropping patterns. A reallocation of water in the LBP from the head-reach to the tail-end would have the same effect, as well as a redistribution of irrigation water use away from the Old Ayacuts and to the LBP.

Finally, the payments made for water should be described. The Revenue Department levied a land tax based both on soil fertility and whether or not the land was irrigated This ranged from 2.5 to 9 rupees (Rs.) per hectare. In the LBP most land was registered at 9 Rs./ha. The land tax was multiplied by water charges based on the crop grown, the multiplier varying between 5 and 50.

For several reasons the land/crop tax had no great impact either on the individual farmer's choice of which crop to cultivate or on the volume of water used for each specific crop. The tax was not a volumetric price. It was low in comparison with the cost of other inputs, such as seeds, fertilizer and labour. Moreover, the collection of the tax was far from complete. In part this was the result of an unclear division of responsibilities and lack of coordination between the Public Works Department and the Revenue Department with respect to crop restriction enforcement. In the late 1980s hardly 10 per cent of the cost of supplying irrigated water was recovered from the users. In 1992 the State government decided to raise crop taxes by more than 300 per cent in order to increase state income and provide a stronger incentive for farmers to grow less water-intensive crops. Irrigated dry crops such as cotton, groundnut and maize were now taxed at 355 Rs./ha for their assumed water requirement of 5,000 m^3/ha; paddy was taxed at 505 Rs./ha for its 12,500 m^3/ha; and sugar cane 655 Rs./ha for its 25,000 m^3/ha.

The LBP Agriculturist Association argued that the tax increase had not been properly advertised and therefore urged farmers only to pay the old tax rates. The case was taken to court and a stay order issued to the effect that members of the Association should only pay the old tax rates until the case had been settled.

In 1996 farmers on the Lower Bhavani Project still based what tax payments they did make on the old rates. Anna Blomqvist's research demonstrates how interdependent are farmers in the same catchment and how powerful are water's externalities, particularly the impact of upstream actions of water appropriation or pollution on downstream users. To reconcile the interests of a group of farmers requires either a state administration of the common resource which is honest, properly funded and competent, or an alternative, farmer-administered institution with similar attributes. By 1996 it appeared that the Lower Bhavani Project enjoyed neither. Other cases of irrigation management transfer will be considered later in this book with more optimistic outcomes.

Reworking the Cycle

The objective of the first case study in the book was simply to apply the concept of the "irrigation cycle" to a real-life situation. A simplistic reconstruction of the "hydrosocial cycle" was introduced in *Economics of Water Resources: An International Perspective.* That first draft proved to be utterly inadequate for the Tamil Nadu case study.

- The growth of the base supply boxes to encompass five categories of supply.
- The strong focus on the constituent elements of drainage both because of their quantitative importance at the catchment level and because they constitute an important source of saline pollution.

In retrospect the Lower Bhavani Project proves to be a rather classic case of a big infrastructure, public-sector system where the storage structure is at the first point of the cycle and where diversions from the reservoir gravity-feed a main canal and a complex of secondary and tertiary channels. The command area of the four main canals below the barrage in total measures some 100,000 ha.

Chapter 2

The Demand for Irrigation Services

CROP WATER REQUIREMENTS

The fundamental concepts of institutional economics may be said to be demand and supply. So it is reasonable to begin building the analytic framework of this book with farmers' demand for irrigation services. But who are these "farmers"? Here it is worthwhile drawing a distinction between farming families and agribusinesses. With the former, a family is at the core of the labour process in crop production. That farming family may or may not engage wage-paid employees, it may grow much of its own food as well as pursuing commercial farming and the size of its landholding is usually small.

In contrast an agribusiness has no family at the core of the labour process, virtually all its staff are employees, little or no crop output is grown which is not sold and holdings are usually large. In this book, for convenience, we shall usually adopt the term farmer, referring to farming families and agribusinesses only where the specific context requires it. Similarly, "farmer" will be applied in the case of share tenants who may be moved by their landlord from time to time from one plot to another.

We start with a community of farmers in a river basin, each of whom has a given hectarage of land with a known cropping pattern. In output terms the farmer seeks tonnage of the crop, product quality and crop composition and timing of the harvest appropriate to the demands of the market. On the demand side, a fundamental agronomic fact is that crop growth depends on water availability to each plant's rootzone.

Soil water generated by precipitation may source this. But in those cases where soil water is known to be insufficient for crop needs between planting and harvesting, then irrigation is required. The same holds true where precipitation and soil water availability cannot be predicted with reasonable accuracy; in this case irrigation sources are necessary to make up for any deficiency that does occur, for example, during a drought period. Where rainfall follows a seasonal cycle, as in the case study of the Lower Bhavani Project the demand for irrigation water is counter-cyclical.

Full irrigation is required when no crop can be reasonably grown unless water is supplied by human effort. It is necessary for virtually any productive agriculture in arid and semi-arid regions. *Supplementary* irrigation indicates a supply of water to crops to increase yields and to reduce the risks of crop failure where, under normal circumstances, reasonable yields can be expected.

Crop water needs, whether met by precipitation or by irrigation, are of central interest to the agronomist. These needs and those for irrigation water can be expressed by a cluster of terms. *Crop water requirements* can be defined as the depth of water needed to meet the evapotranspiration requirements of a disease-free crop, growing in large fields under non-restricting soil conditions and achieving full production potential under the given growing environment. Alongside this optimum volume we have *actual crop evapotranspiration*. Next comes *effective precipitation* defined as that part of total rainfall that can be beneficially used by crops. Lastly, we have *net irrigation requirements*, the amount of irrigation water needed to supplement rainfall to meet crop water requirement, excluding irrigation water lost to nonbeneficial evapotranspiration and drainage.

Different crops have different crop water requirements. There the irrigated "dry crops" were cotton, groundnuts and maize, whereas sugar-cane was expected to consume five times as much water per hectare. When it is feasible, farmers tend to grow dry crops in the dry season and wet crops in the wet season. Actual evapotranspiration of a typical crop is about 500 mm (5,000 m^3/ha) between planting and harvesting.

In this book we shall repeatedly use various measures of irrigation efficiency and productivity, terms used interchangeably. The first of these, E_1, is an output to input ratio known as *relative efficiency*. Hargreaves and Christiansen (1974) have shown that E_1 is strikingly similar across crops. Their data derived from Davis, California (corn and alfalfa), Hawaii (sugar-cane) and Delhi, India (alfalfa). Here the input is measured as actual water supplied divided by the physical optimum for plant growth and output is measured as actual crop tonnage divided by the physical maximum, with all other factors such as seed and fertilizer optimized.

E_1 is dimensionless, a pure number. Between the values of about 15 per cent and 65 per cent of the water input variable, the relative efficiency function is virtually linear; here a ten-point increase in water supply is matched by about a ten-point gain in output. Beyond about 65 per cent, output gain from a one-point water supply increase begins to weaken and, by definition, it becomes negative beyond the E_1 value of 1.0. Plants can be supplied with too much as well as too little water for their maximum growth. (Exceptionally, rice has the capacity to grow in saturated conditions.) Overwatering may take place as a result of uncertainties in irrigation distribution. In the reverse case, when water not land is the limiting input, farmers may spread their water over a larger crop area.

How important, on a world scale, is irrigation in the supply of water for agriculture? van Hofwegen and Svendsen distinguish between four water management regimes: rain-fed, irrigation only, drainage only and irrigation and drainage. They write:

> Of the 1,500 million hectares of global cropland some 250 million hectares (17%) are irrigated and approximately 150 million hectares are provided with drainage infrastructure. Currently, about 60% of global food crop production originates from rainfed agriculture and the remaining 40% from irrigated agriculture.

DETERMINANTS OF WATER EFFICIENCY

Water efficiency, in terms of these physical measures, has many determinants other than the volume of water applied at the rootzone. One of these concerns the timing of rainfall or plant irrigation. Crop water requirements vary across the season from planting to harvesting. For example, water stress at the flowering stage of maize reduces yields by 60 per cent, even if water is adequate during the rest of the crop season. Similarly, the need for water can be critical at seeding time for germination and when seedlings are transplanted to the field. For these reasons cultivators need to maintain adequate control over irrigation supply so that, in the ideal situation, water inputs are available precisely when they are needed.

The desire for precision in the timing, volume and physical placement of irrigation water has stimulated water-use technologies that promise to achieve this, particularly sprinkler, drip and trickle irrigation. In practice, without adequate management, the new technologies can be less efficient than simpler surface methods.

Another determinant of water efficiency is the volume and quality of farmers' other inputs to their cultivation practices. Better seeds, the right quantity of the right fertilizer, biocides to reduce crop losses and plentiful skilled labour at the right time of the season all boost field and basin water efficiencies by boosting output volume. Moreover, it has been the availability of irrigation services that has provided the historical platform for these other contributions to water efficiency. Irrigation reduced the risk that farmers faced in rain-fed agriculture with its intermittent drought and subsequent crop failures. This manufacture of risk *reduction* made the decision to spend on other inputs much more profitable. For example, the extraordinary advances in volume of output, labour productivity and sales in the Indian fertilizer industry from the mid-1960s was certainly underpinned by growth in the irrigated area:

$$E_2 = \left(\frac{N}{S_b}\right)$$

where E_2 is requirements efficiency, N is the net irrigation requirements and S_b is defined above. The difference between N and S_b is equal to the storage and distribution losses between source and field, non-beneficial evaporation at the field level and field drainage. Note that E_1, *relative* efficiency, deals with a relative ratio between crop output and water use, whereas requirements efficiency deals with an *absolute* ratio between evapotranspiration and water supply. As with E_1, E_2 is also a pure number.

The ratio of N to S_b can be expressed in another way. Where water efficiency uses as its input measure the gross irrigation volume drawn off by rainfall collection, surface and groundwater abstraction, reused wastewater and reused drainage water, then the more of this volume that is *captured* for the crops' net irrigation needs the greater will be the measured water efficiency E_2. This is of tremendous importance in increasing "crop per drop".

It should be noted that the measurements required to estimate requirements efficiency are extremely demanding, including as they do crop water requirements, the irrigation water quantity needed to supplement rainfall at the plant's rootzone and the multifarious losses between the supply source of the irrigation flow and the rootzone. Here, as in other ways, the nature of the knowledge developed and applied by professional economists and agronomists, for example, and that of the farmer is likely to differ considerably. In my view the primary interest of the economist in understanding the irrigation and drainage sector should be the values, beliefs, motives and behavioural routines of the sector's actors, amongst whom farmers are the protagonists. Only on this basis can an effective economics of irrigation be constructed.

THE COST OF WATER TO FARMERS

Physical measures of water efficiency can be of value to water resource planners when they draw up a scenario hydrosocial balance for the purposes of basin strategy formulation. For example, we can write as *tonnage efficiency:*

$$E_3 = \left(\frac{T}{S_b}\right)$$

where T is the actual tonnage of crop output in a region in the baseline year and S_b is the volume of irrigation water supplied at the regional level. In this case estimates of future desired crop output could be divided by tonnage efficiency to derive a first estimate of required irrigation supply in agriculture in millions of cubic metres in the scenario year. But water efficiency measurement in this way may be of little account in planning by farming families or agribusinesses. Their behaviour is shaped above all by economic motives and therefore by cash costs of inputs and value of outputs, not by the ratio of tonnage to water quantity ratios of agronomy and resource planning. So it is necessary to consider what are *the cash costs to farmers of their use of water*.

The *first case* is the simplest. A public-sector irrigation authority provides water free (or virtually free) of cost. Malaysia is an example; here, a nominal charge is made for water in the annual fixed land tax payable by farmers. In 1997 this was 15 Malaysian ringitts, or about US$0.50, per month. In this limiting case the irrigation flows are treated by the farmer as a free resource, no more expensive than rainfall. As an economic unit the farming family or agribusiness regards the water as simply a critical natural resource that, with precipitation, determines output.

The *second case* occurs where the farmer provides his own supply of water through rainfall collection, appropriating surface water or pumping groundwater. But it gives nothing away to report that, for example, a family with its own tubewell must bear its installation and running costs. The family's cash outlays for its water supply are here very real, particularly if the price of energy is high. In this example water can be measured not only in cubic metres but also in Indian rupees, Chinese renminbi, Pakistani rupees, US dollars, Mexican new pesos or what you will. Note that in this situation farmers also deploy their labour time in supplying themselves with water-the traditional Egyptian shadouf is a ready example. So there is a labour cost not expressed in cash but with a significant opportunity cost. In this case the opportunity cost is the output value foregone or the leisure time sacrificed by the farmer as a result of committing time to supplying the farm with irrigation water.

The *third case* in considering the cash costs of water to the farmer is where a volumetric price is charged by an irrigation authority. Here the uptake of water by the farm is measured in volume terms and a price per cubic metre (for example) is charged. (The water tariff per unit may vary with the total volume used and with the time of day or the season during which the water is supplied.) It is pricing of this kind, for goods and services in general, that is the basis of most academic microeconomic analysis in market societies.

The *fourth case* is where an irrigation authority, or its equivalent such as a water users' association, makes a charge on the farming family or agribusiness that correlates with its water consumption per hectare but is not in the form of a volumetric price as in the third case. In this fourth case the charge paid by the farmer varies with the volume of water used but in an *indirect* way. The Department also used a multiplier in calculating the final tax bill to irrigators of 355 Rs./ha for the presumed water requirements of 5,000 m^3/ha for cotton and 505 Rs./ha for the presumed water requirements of 12,500 m^3/ha for paddy.

Another example of the fourth case situation is where farmers pay cash for their supplies of irrigation water on the basis of the length of time for which they have access to that supply. Julie Trottier has a case study of Jericho's Ein Sultan spring that illustrates this situation. If the flow of irrigation water per unit time did not vary, a time-allocation price and volumetric pricing would give the same outcome. But, of course, irrigation water flows *do* vary with time. Time allocation is also used by the Mexican water user associations. The best known of all time-allocation arrangements is the *warabundi* method of rotation described by Jeremy Berkoff for the North India Act systems. Water is allocated in proportion to land and water deliveries are controlled by time; full flow in the watercourse being used by each farmer in turn.

The *fifth and last case* is where an irrigation service fee is charged by the irrigation authority but the payment is for revenue-raising purposes alone and bears no relation, direct or indirect, to the volume of water used.

From the point of view of water resource planners, in those cases where a cost-tool can influence farm use of water, it can be deployed to raise water efficiency. Where it contributes to an irrigation authority's revenue, it can be deployed to finance all or part of that authority's capital and operational costs. Only the third and fourth cases satisfy both criteria.

THE ELASTICITY OF DEMAND

Volumetric price and indirect fee may all influence the demand for irrigation water by farming families and agribusinesses. Economists measure the responsiveness of volume demanded to price change by the price elasticity of demand.

The vertical axis on the left-hand side measures *P*, the price of irrigation water as in the volumetric fee case and the horizontal axis measures *Q*, the amount of irrigation water purchased by the farm each month. Embodies four assumptions. *First,* there exists a price per unit of irrigation water so high that the volume purchased is zero; conversely, when the price of water is zero, there exists a defined highest volume purchased.

Second, throughout its length the demand function slopes downwards from left to right; a lower price is always associated with a higher volume purchased. *Third,* the units in which price and quantity are measured have been chosen so that at price 100 no water is purchased, whereas 100 units of water are purchased at price zero. This standardizing device can be used for any true-life situation. *Fourth,* the slope of the demand function is linear. This form is the simplest option for the figure. Other alternatives are curvilinear functions (convex or concave to the origin) and a cubic function.

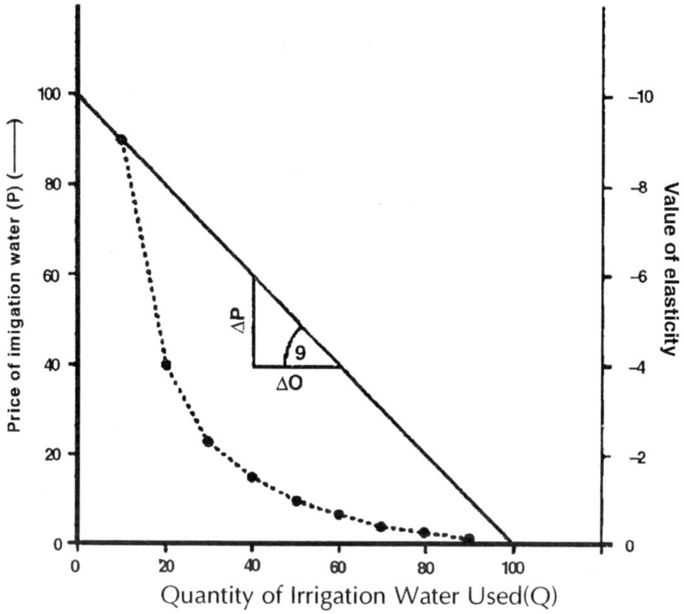

Fig:1. The Demand for Irrigation Water

The price elasticity of demand is defined as the proportionate change in the quantity demanded of a product divided by the proportionate change in its price:

$$e = \left(\frac{\Delta Q/Q}{\Delta P/P}\right)$$

where e is elasticity and ÄQ and ÄP are small changes in the value of quantity and price. (lower-case e is conventionally used as the symbol for elasticity, whereas we use the upper case E to refer to water efficiency.)

One can write the slope or gradient (g) of a demand function as:

$$g = \frac{\Delta P}{\Delta Q}$$

This allows the elasticity variable to be re-written as a function of the gradient and the P-Q pair at the point of measurement:

$$e = \frac{P}{(gQ)}$$

From assumption to above, g (therefore e) is negative. The value of g

is "1 for the standardized form of the linear function. Therefore we can write:

$$e = -\left(\frac{P}{Q}\right)$$

So the lower the ratio of price to quantity at any point on the demand function, the lower is the absolute value of the elasticity. The value of *e* shifts from "9.0 at the *P*-*Q* pair (90, 10) through "1.0 at (50, 50) to "0.1 at (10, 90). The elasticity values for the linear demand case at intervals of five units between *Q*=5 and *Q*=95. This table also sets out the values of the elasticity function for an arc demand case (strongly) convex to the origin, a common representation of the demand curve for water. Whenever volumetric pricing or own-supply costs operate and there are no restrictions on the volume of irrigation water used by the farmer, has a practical application in the field for agricultural and irrigation economists. In such a situation the economist can:

- Take a view as to whether the demand for irrigation water takes the linear or the arc form.
- Estimate, for the volume of water already used, what percentage that is of the maximum volume that would be used were the cost of water to be zero.
- The elasticity of demand in the current situation for the chosen function type and the estimated *Q* percentage.

It must be admitted that the number of convex functions is infinite. Our suggestion is that the economist in the field should use the linear function elasticities unless s/he has good reason to believe an alternative curvilinear function should be preferred. *Note that a low price (or cost) per unit of irrigation water is very strongly associated with low elasticity in the case of both the linear and convex functions.*

Farm Budgets

The purpose is to understand what water efficiency in economic terms may mean to cultivators and how responsive they may be to changes in a volumetric price or, indeed, how they might respond to the *introduction* of volumetric pricing. The way in which a farmer constructs the farm budget varies between farms, between catchments and between cultures. The approach is also likely to be different between farming families and agribusinesses. This is of no great importance here, where the purpose is to tabulate in a down-to-earth way the principal economic variables impinging on a cultivator's life.

- *Prime costs* are defined as expenditure on those inputs used up in the daily round of farm work. They consist both of the salaries and wages of the workforce's labour time as well as the costs of electricity, fertilizers, other agrochemicals, seed, fuel for farm machinery or vehicles, spare parts, irrigation water and other consumables such as bought-in specialist inputs and services.

- *Overhead costs* of production are non-prime costs and are often assumed to be invariant with the level of crop output. They consist of the rent of land and buildings, insurance, taxes, loan interest and debt repayment on borrowings. Where expenditure on farm machinery, for example, has not been funded by a loan but out of the accrued profit of an agribusiness, the farm may include in its budget the amortization charges necessary in the long run to replace the plant when it is scrapped.
- *Total costs* per year are equal to prime plus overhead costs.

Income is derived, in this case, from both irrigated and non-irrigated crops, as well as any other sales of farm produce. The farm's gross margin per hectare is the excess of total income over total prime costs, divided by farm size and is the source of income to cover the farm's overheads; it is often referred to as "net return". Total income less total costs gives the farming family or agribusiness its annual surplus. In this book the terms "income" and "turnover" will be used interchangeably.

The farmer's demand for water is itself derived from the market's demand for the farmer's outputs. In terms of economic theory the farmer may estimate for irrigated crop output a rather sophisticated water efficiency measure, the increase in farm tonnage following a modest addition to the supply of irrigation water, i.e., *marginal tonnage efficiency* (E_4). (Neoclassical economists call this the marginal productivity of water, which can be defined in terms, for example, of base supply or field supply as appropriate.) Using their knowledge of how total income to the farm would increase with the additional output and how the prime costs of water would rise, farmers would have a marginal efficiency measure *in economic terms* (E_5). E_5 transposes E_4 from physical to cash terms and I shall call it *marginal turnover efficiency*. From this perspective, an increase in the price of water would probably be followed by a reduction in the demand for water. Here other inputs are assumed to be unchanged. Similarly, other things being equal and for a known value of the net irrigation requirement, the greater the gross margin of any single crop, the greater will be the quantity of irrigation water purchased for its cultivation at any given price of water. Conversely, when the farmer makes choices between the crops to be grown, the higher the price of water, the less likely it is that crops with low values of E_3, E_4 or E_5 will be selected. Which one of these three efficiencies is used by farmers doubtless varies; in economic terms, E_5 is the preferred criterion.

Economic theory is by no means always close to farmer behaviour in this respect. *First,* such rational action is most effective only in the cases of ownsupply cost and volumetric price. *Second,* it is feasible only where the farmer has control over the volume of water he can apply to the land. *Third,* the farmer may not know the response of crop output to marginal change in water use, adopting a satisficing rather than an optimizing pattern of behaviour. *Fourth,* if the price of water is low or if the proportion that water outlays make up in total prime costs is small, the marginal adjustment of this input may be seen as unimportant so the elasticity of demand for irrigation water would be negligibly low. Here, in respect of prime costs, irrigation water is not an economic commodity to the farmer but is simply an agronomic necessity.

To conclude, in all catchments and for all families and agribusinesses, water is certainly seen as a vital natural resource in crop production but is less often viewed as an economic input. In the latter case, it will not be sensitive to price manipulation by a catchment agency in the management of demand.

THE MANAGEMENT OF DEMAND

The preceding paragraph refers to "the management of demand". This term has entered the mainstream of hydroeconomics in the last decade or so and requires an explanation of its full meaning in the context of water resource planning.

An important debate in the water industry in recent years has concerned the forecasting of water use, usually referred to as "demand forecasts". The approach to forecasting in the 1960s and thereafter has been strongly criticized, as have supply-fix philosophies. Objective forecasting techniques take two forms. With *extrapolation* forecasts, past levels of aggregate use are recorded and their rate of growth is projected into the future. With *component* forecasts (also known as the causal or analytical technique), the different categories of water use are identified and their rates of growth into the future are estimated with demographic and economic projections relevant to each category. In this second case, there can be heavy data requirements. Perhaps for this reason, extrapolation is more commonly used.

The critique of forecasting method is wide-ranging, particularly in respect of extrapolation. The arguments are that it ignores the components of use, assumes linear growth into the future, often based on the steepest sections of the trend line and adds overgenerous safety margins against unpredicted surges in use. Component forecasts are innocent of the first of these charges, since this technique *bases* itself on population growth and predicted industrial and commercial development.

However, both techniques are guilty of two further methodological errors. One is the misidentification of supply losses as a component of use. The second error is to regard use as a simple technical issue, exogenous to other social changes. It is precisely these changes that are at the heart of what has become known as demand management, a concept that has at least six strands: internal and external reuse, use technology, land-use planning, environmental education, water pricing and food import policies. With internal reuse, a specific consumer first uses the water supplied to it, such as from a water service company, then returns its wastewater internally for a second round of use and then perhaps a third and so on. For example, the installation of water storage during housing construction can make the one-off reuse of bath and shower water for flush-toilets perfectly feasible, reducing total household water use by perhaps 15 per cent.

With external reuse, a consumer, whether household, farmer or industrialist, uses the water supply and then the waste or drainage water is supplied as an input to another institution. With both internal and external reuse, the wastewater often requires treatment. Reuse engineering is best interpreted as a supply-side innovation. At the same time, it brings about a lower aggregate demand for water by the community as a whole than would happen in the absence of reuse and, in that sense, reuse is also treated as a form of demand management.

Use technology, the second form of demand management, is similar to internal reuse, in that both deploy technical measures that reduce take-up of the external water supply. In this case, in respect of households, the redesign of showers, toilet cisterns, washing machines and dishwashers can cut water use significantly. Household access to the new technology may be through the purchase of new machines or in retrofit programmes of the type launched in some US municipalities. Household acceptance of such innovations can be underpinned by government through building regulations and water by-laws.

In the case of irrigation, McCann and Appleton (1993:38) have pointed out that in France, for example, private abstraction is permitted by land ownership and past practice. It is a prime source of water wastage because of inefficient spraying devices and methods and just straightforward overspraying. The wasteful use of water in industry is also widespread in Europe, in the view of informed industrial opinion.

Land-use planning is the third form of demand management. The argument concerns catchments where use is pushing at the very limits of supply capacity, or threatens soon to do so and where increased abstraction would impose relatively high economic or environmental costs. In such cases, as Gordon and Merrett argue, land-use planning can restrain urban development and the consumption that accompanies it–and divert its location to other regions not facing the same supply/consumption imbalances.

The fourth form of demand management is to use *environmental education* to persuade the population, as citizens, farmers and managers in industry, to use water wisely. Many US municipalities, for example, provide comprehensive water-use advice and information services, including residential water audits. Trained personnel inspect water use fittings and appliances in individual homes, undertake leak detection and repair, evaluate lawn watering practices, advise on low-water garden design and recommend water-saving plants".

US studies suggest that engineering audits, technical workshops and best-practice manuals can save 20-35 per cent of total annual water use with a short average payback period. Educational measures bring about a change in consumers' tastes and habits. Most orthodox economics takes preference functions as being fixed.

A fifth form of demand management is *water pricing*, bringing us back to the demand for water in the strict economic sense of a price-quantity function. The metering of water use and setting a price per unit quantity, helps to underpin all the first four demand management forms. Water pricing should therefore be considered as complementary to these other measures, not as a substitute for them.

In many parts of the urban world, no price is paid for water. In the case of households, the reason can be found in the traditions of the social services approach. In this case it is accepted that the supply of water should be regarded as a municipal service. Wasteful use, stimulated by the zero price, is then built into demand forecasts of the extrapolative and component type.

However, the growth in the economic costs of treating potable water, the environmental costs of abstraction, the privatization of the water industry and the spectre of water closure have all brought metering and pricing onto political agendas. In the UK in the year 2000, for example, 82 per cent of households lived in accommodation where their water consumption was unmetered and where the costs of water were met not by a price for its use but by a property-based fixed charge levied by the private water service companies. The key difference in revenue raising between pricing and a fixed charge is that only the former is based on the quantity used by the household.

Where prices are to be introduced for consumers, price-setting policy has to be considered. Water price tariffs come in many varieties. There is likely to be a fixed standing charge for connection to the service, irrespective of total use. The unit price of water consumed may be on a flat rate basis, or may increase (or fall) with each successive block of consumption. Higher prices at consumption peaks may be put in place. The peaking of water use may be seasonal or in the so-called "needle peaks" at peak hours. User peaking, to be satisfied on the supply side, requires a greater system capacity than in the absence of peaks and "average day in peak week" forecasts are conventionally employed in water investment planning. So, demand-side peak smoothing, in a long-term perspective, can offer considerable cost savings. For this reason, higher seasonal and needle tariffs are particularly attractive.

Finally, *food import policies* can play a very substantial part in management of the demand for water, particularly in semi-arid and arid countries. If a national government takes the decision to restrain the growth or cut back the size of the agricultural sector and to place a growing reliance on imports to feed the domestic population, the demand for irrigation water will be less than in the no-policy-change situation.

THE DEMAND FOR IRRIGATION WATER

The material is derived from Karin Kemper's detailed and perceptive research carried out in 1993-96 and published as The Cost of Free Water: Water Resources Allocation and Use in the Curu Valley. Ceara is one of the federal states of Brazil and is located in the semi-arid north-east. The area is chronically short of water and over the past century federal and state agencies as well as the private sector have tried to solve this problem by building dams to catch water during the rainy season for distribution in the dry season and in the recurrent drought years. The Curu Basin covers an area of 7,900 km^2 and extends for about 160 km from south-east to north-east where the river flows into the Atlantic. The irrigated area follows the humid valley and is located between General Sampaio reservoir and the ocean. Along this 120-km stretch, average annual precipitation varies from 750 to 1,200 mm. Evaporation, measured at the dams, exceeds 2,000 mm. The rainy season is between January and June with no precipitation during the rest of the year. There are strong inter-annual variations in precipitation. Maximum average temperature varies between 31°C and 35°C over the year. The hydrogeology of the area with its crystalline rocks does not favour groundwater as a resource. A good road infrastructure and proximity to the state capital of Fortaleza with its population of 2.4 million permit ready access to urban markets.

Irrigators fall into four groups. *First,* there are 300 families and small agribusinesses that are "private" farmers found along the river, each with an irrigation area of some 2-10 ha and their main crops are banana, beans, foddergrass and maize, with sugar cane in the downstream area. *Second,* there are about 200 "project" farming families whose plots are located on an irrigation project, known as Curu-Recuperação, implemented by DNOCS, the Brazilian National Department of Works against Droughts and they have 3-4 ha each. They are fed water by gravity from the General Sampaio and Serrota dams and their main crops are banana, beans, coconuts and foddergrass. *Third,* there are about 800 families and small agribusinesses on the DNOCS Curu-Paraipaba project, located on a plateau and they receive water pumped from the river by electric motors, the main crops being coconuts, foddergrass, manioc and sugar cane. *Fourth,* there are three large-scale agribusinesses. Agrovale and Ypióca specialize in sugarcane cultivation for alcohol and spirits and pump water directly from the river. The third company, FAISA, receives its water from the DNOCS's main canal to grow acerola and has a separate acerola and melon plantation using pumped river water. Of about 170 mcm of water used in the Valley in 1995, 90 per cent was for irrigation. The irrigated area was about 7,800 ha.

The share of different technologies is 52 per cent for sprinklers, 33 per cent for flood irrigation, 11 per cent for gravity-fed channels and only 4 per cent for drip-virtually all of it on FAISA's acerola plantation. A consultancy report for the Valley suggests benchmark values of field efficiency at 70 per cent for sprinklers, 60 per cent for gravity and 90 per cent for drip irrigation. Actual field efficiency is estimated as 65 per cent for sprinklers but only 40 per cent for gravity systems. The consultancy report had been intended to provide a basis for estimating water use in the Valley. However, "in spite of the efforts undertaken, the attempt did not show any success whatsoever due to the complete ignorance by the consumers themselves concerning the amount of water they use".

The cost of water to farmers in the Curu Valley is considered next, using the categories of the specific truths of the Valley in the mid-1990s. It should be noted that while nobody in the Curu Valley has quantified formal abstraction rights, Kemper's interviews demonstrated that "all respondents are of the opinion that they do have the right to draw as much water as they perceive they need".

Group 1, "private" farming families along the river, are not required to pay volumetric prices, indirect fees or revenue-only fees for their water. That is to say, these irrigation users do not pay for any of the supply costs of the infrastructure provided by federal and state government. However, this group irrigates with water abstracted from the Curu River by electric motors. Clearly, the more water pumped, the greater the amount of electricity consumed. Electricity consumers are billed by COELCE, Ceará's publicly owned power utility. Their collection rate is well above 90 per cent. COELCE divides its consumers into two categories: high-tension consumers (using motors of more than 50 kW) and low-tension consumers.

"Private" farming families, as low-tension customers, pay no fixed charge and in 1994 faced an unvarying unit tariff of US$0.064/kWh. Some evidence exists from projects in Ceará financed by the Kreditanstalt für Wiederaufbau that such costs do influence water consumption.

Group 2, "project" families with DNOC's Curu-Recuperação project, are gravity-fed and so face no electricity bill. Nor do they pay a volumetric price or an indirect fee. Their outlays are limited to two revenue-only fees. The first, K_1, is paid monthly and is related to the lease of the land and infrastructural amortization. The second, K_2, is based on project personnel costs, DNOCS personnel costs, energy costs for irrigation, project maintenance costs, maintenance costs for motor vehicles and administrative expenses, less a DNOCS subsidy. The manager of the project calculates K_2 for the whole project, systematically understating it in the process and divides it by the number of irrigators so that each pays the same amount, irrespective of how much water is used and which crop is grown. The K_2 payments stay within the project.

The rationale is that the irrigators would not be able to pay the real costs anyway since they are too poor. It is needless to say that, without other sources of income, the continuous underreporting of actual costs-and the consequent scarcity of financial resources for operation and maintenance-leads to a progressive deterioration of the project. In addition, this type of reasoning perpetuates the general situation in the project...the project is quite paralysed.

Group 3, "project" families with DNOC's Curu-Paraipaba project, also pay a bill of the K_2 type above, but in their case it includes the electricity costs of irrigation pumping and this component, at least, might be regarded as an own-supply cost. However, here the project's internal charge system comes into play. House electricity connections are individually metered; consumption at plot level is not. COELCE sends a lump sum bill to the district manager, who divides it by the number of irrigators, incorporates it into K_2 and passes it on to each family. Given the large number of irrigators, it is clear that, if one family decided to use water more efficiently in order to reduce its electricity bill, the cash saving would be close to zero for that family. So the K_2 bill at Curu-Paraipaba is also a revenue-only fee. Group 4, the agribusinesses, are high-tension customers and they pay a fixed charge plus a tariff based on the amount of energy they use. However, they all qualify for a 50 per cent reduction of the price because they commit themselves not to use energy in the peak hours from 17.30 to 20.30. Moreover, they receive a 90 per cent reduction of the conventional tariff when they pump between 23.00 and 05.00. Agrovale and Ypióca both take advantage of this discount.

The two large private consumers do respond to economic instruments.

Of course, the fact that they shift their pumping of water to other times of the day does not conserve any water, only energy. The differentiated energy tariffs do not decrease water consumption in the Valley. Rather, they can be expected to increase consumption since it means that for six hours during the night pumping water is almost free.

The third agribusiness, FAISA, is said to have installed the so-called xique-xique irrigation system for its acerola plantation because it was the cheapest from an investment point of view. The energy costs of water were not an issue. Kemper states that the intention of the government to subsidize irrigation activity by providing cheap energy and thus cheap water, is absolutely clear.

In summary, water charges in the Curu Valley are of a revenue-only type, except for the case of off-project irrigation requiring pumping, where they are of the own-supply type with respect to electricity use. Even in this case, the energy costs of water to the three large-scale agribusinesses play no part in their decisions about water use. This Brazilian case study ends by taking up an issue that I have touched on only briefly -cropping pattern. However, my interest here is limited to the relationship between cropping pattern and the economics of irrigation, not to the much broader issue of crop choice as it is dealt with in texts on the economics of agriculture as a whole. From a theoretical point of view the strongest argument for an economic influence on cropping pattern derived from the money costs or fees paid for irrigation water is where:

- There are marked inter-crop differences in E_5, i.e., marginal turnover efficiency.
- The cost of water to the irrigator falls into one or other of three categories: own-supply cost, volumetric fee or indirect fee.
- The cost of water is significant as a proportion of the prime costs.

Since all the Curu Valley evidence points to very weak compliance with conditions 2 and 3 above, theory suggests that E_5 would be a poor predictor of crop composition there.

The crops (save manioc) with at least 1 per cent of irrigated hectarage in the Valley, gives data on the scale of the irrigated area by crop and its assessed water efficiency. The ranking in these two variables shows that there *is* a loose association between hectarage and efficiency-but with the wrong sign. The higher the efficiency, the less land is allocated to the crop! So we can safely say that a crop's strong performance in water efficiency does not encourage its scale of cultivation compared with other crops. Theory stands confirmed. Note that in column 3 we now have a sixth E-value that we shall call *gross margin efficiency:*

$$E_6 = \frac{G}{S_b}$$

where *G* is equal to the farm's gross margin per hectare in US dollars and S_b is as already defined but on a per hectare basis. This is a dollar-water quantity ratio, the hectares in numerator and denominator being self-cancelling. For the gross margin efficiency and all the other efficiency measures.

In practice the explanation of crop composition in the Curu Valley in the mid-1990s had nothing to do with the economic demand for water but was related to other issues. Small farming families favour beans, manioc and maize as subsistence crops. Agrovale and Ypióca are manufacturing companies growing sugar cane for the alcohol they produce.

Table:1. Irrigated Crops in the Curu Valley: Mid-1990s

Crop	*Irrigated Area (ha)*	*Water Efficiency (US cents/m3/ha)*[a]	*Area Rank*	*Efficiency Rank*
Sugarcane	3,369	5	1	10
Beans	1,455	10	2	7
Foddergrass	917	8	3	9
Coconut	878	10	4	8
Banana	359	21	5	4
Acerola	326	27	6	3
Maize	294	16	7	5
Papaya	189	41	8	1
Pumpkin	71	13	9	6
Citrus	69	28	10	2

Together these two agribusinesses control about one-third of the irrigated hectarage in the Valley. Foddergrass is used to feed families' cattle, rather than being sold and is another example of a crop grown as an input to a second economic activity rather than for sale. Coconut and papaya need significant investment before they yield any return. Drought-sensitive crops tend to be grown on the upstream stretches, as water security is greater there. Finally, different types of market and marketing channels have a powerful influence on irrigators' crop choice.

IRRIGATION WATER USE AT KEVERAL FARM

Keveral Farm is located close to Seaton on the low hills rising from the south coast of Cornwall in the UK. The farm is worked and managed by some seven or eight persons and is legally constituted as a cooperative company limited by guarantee known as Keveral Farmers Ltd. This group of cooperators sits neatly between the categories of farming family and agribusiness. The farm has 30 acres (12.2 ha), of which just 5 acres (2.0 ha) are irrigated. But the number of crops per year varies from one to five depending on the product, so that the gross irrigated acreage is about 12 acres (4.9 ha). Gross irrigation area is equal to the net irrigation area (i.e., the actual irrigation area) multiplied by the average number of times that crops are grown per year on it. An actual irrigation area of 1 acre (0.4 ha) would be counted as a gross irrigation area of 2 acres (0.81 ha) if it were cropped twice in the year.

Keveral Farmers Lt mainly produces vegetables and soft fruits: beetroot, carrots, leeks, onions, parsnips, potatoes, swedes, turnips, broccoli, cabbage, cauliflower, kale, oriental brassicas, spinach, asparagus, aubergine, beans, broadbeans, courgettes, cucumber, custard marrow, herbs, lettuce, mushrooms, peas, peppers, tomatoes, strawberries and raspberries. All the produce is grown organically and the farm is managed following the aims and principles of the Soil Association, which certificates Keveral as qualifying as organic under the Association's demanding criteria. All the crops grown are irrigated.

About 3 per cent of output meets the personal needs of the co-operators; some goes to cafés and market stalls; and about 90 per cent is sold through the "organic veggie boxes" scheme. Under this arrangement about 180 boxes of fresh, organic produce from the farm is packed each week and delivered by the cooperative to meet the orders of individual households either at their homes or at common drop-off points, such as the Environment Agency's offices in Bodmin. These deliveries are made to the villages and small towns close to the farm such as Looe, Downderry, Sheviock and Morval.

Product quality is critical–in terms of freshness and organic sourcing–and purchasers are aware that this is rarely compatible with the cosmetic appearance of supermarket sales of agrochemically based outputs. Boxes come in three prices: £5, £7 and £9. The farm has to meet the transport costs of their distribution network but receives 100 per cent of the retail price paid by the customer. The timing of crop availability is also critical-in this case to maintaining the customer base. Intermittent failure to deliver the weekly veggie boxes would quickly lead to loss of sales and termination of the farm's income stream. Admittedly there is a "hunger gap" each year between late April and early June when little produce comes off the fields. Over the course of the year a substantial proportion of the boxes' contents is made up from other organic sources in Cornwall, Devon and as far away as Egypt and Morocco. Product variety is an important marketing advantage and this accounts for the wide assortment of vegetables grown.

Keveral Farm has four water sources. *The first* is a leat, the Cornish term for a contour ditch and this drains the hillside marking the farm's south- and west-aligned boundaries. Many decades ago an overshot water wheel turned here. The leat drains several small ponds and finally its discharge seeps down to the River Seaton below. The farm produces its own seedlings and they may be started off in the main pond. Otherwise the leat plays no part in the irrigation cycle.

The *second* supply source is a 43-m borehole driven into the sandstone aquifer. This supply point is used for irrigation only in emergencies. The *third* flow is captured by harvesting the rain that falls on the farmhouse. Gutters lead the precipitation to a downpipe that slopes directly into the "new" storage tank. The roof captures 150 m^3 in an average year, when rainfall in this part of Cornwall is some 900 mm. The *fourth* water source comes from a stream that drains the north boundary of the farm. Part of its flow is diverted into three concrete tanks and from here a diesel-engine-powered pump raises the flow to the "old" storage tank at the highest point on the farm, close to the farmhouse. The third and fourth flows dominate the supply side of Keveral Farm's irrigation cycle. Distribution from the two storage tanks to the crops is entirely by gravity.

The co-operator who was my principal informant during the fieldwork in May 2000 did not know how much water in total was supplied to the farm nor, specifically, how much water was supplied for irrigation purposes. None of the four flows referred to above is metered. There is a parallel here with farming families and agribusinesses in the Curu Valley in the previous case study. However, knowing the capacity of the two storage tanks (10 m^3 and 8 m^3) and estimating how many times per week these tanks are emptied for irrigation purposes at different times of the year, my informant calculated an annual rate of supply from storage to the crops of about 600 m^3 : 500 m^3 drawn from the stream and 100 m^3 from rainwater collection. Irrigation at Keveral Farm takes two forms: on open fields and in five polythene greenhouses, one of which is twin-span. The use of irrigation water on the former compensates for periods of dry weather. Irrigation gets under way after any spell of 7-9 days without rain. As indicated above, average precipitation in the area is plentiful and so the irrigation requirements of field crops are modest. Swales are used to reduce surface run-off and soil erosion.

The greenhouses act as a means of climate control at the microscale. Because no rain enters them, crop water requirement is met entirely by irrigation. Given the farm's location, ambient control is a prerequisite to produce the volume, quality and variety of produce that the marketing of organic vegetables requires. The greenhouses limit the temperature range to which the plants are exposed, thereby eliminating this component of farm risk. They create a warm and humid environment that is particularly suitable for the "Mediterranean" crops; it is also a microclimate that checks the depredations of pests such as the red spider mite. The greenhouses shelter the growing plants from the often-blustery winds across the coastal hills-winds that can flatten vegetables or inhibit their growth. Finally, greenhouses keep the rain off! This discourages the entry of hungry legions of snails and slugs, it prevents plants receiving excess water that reduces output quality and it gives the co-operators a small warm world of work when lashing rain makes open-air labour unpleasant and inefficient.

My informant does not know the crop water requirements of any of the farm's produce, at least not in the agronomic sense of being able to cite annual requirement by crop per year in cubic metres per acre. In another, non-statistical sense, he understands the crops' water needs very well-through experience, trial and error and day-to-day observation, such as by thrusting his fingers into the soil to check how damp or dry are the top 2-3 inches (5-7 cm). For example, a crop of tomatoes was receiving irrigation water for 1 h in 24, but yellowing of the leaves due to overwatering was taking place so the irrigation schedule was changed. Too much watering in the greenhouses in the winter months triggers fungal disease. We calculated the actual volume of water use as follows. Outflow from the two main tanks for irrigation was some 600 m^3 in 1999-2000. Losses between the tanks and the fields or greenhouses is small-about 5 per cent. Leaching requirements are zero in organic farming. About 10 per cent of the gross supply probably represents the overwatering associated with introduction of new systems and insufficient staff training. This gives about 540 m^3/year for crop water requirements, about 570 m^3/year in actual evapotranspiration and about 48 m^3/ acre/year of water used in terms of the gross acreage.

A ratio of crop tonnage to water used has no practical meaning at Keveral Farm. Output is calculated in a way that relates to marketing, in weight (or units) per week per crop, such as 300 lbs (135 kg) of carrots/week or 250 lettuces/week. Planning and planting is done on the same basis. I now consider the cost of irrigation water to Keverál Farmers. The *capital* costs of supply are the rainfall collection infrastructure, the diesel engine and pump at the stream, a spare pump, the two storage tanks and the plastic-piped water distribution network to the fields. The *current* costs of own-supply in this case are the co-operators' time in maintenance of the supply network, spare parts and diesel fuel and lubrication oil for the diesel engine.

The only operating costs requiring money payment that have a volumetric relation to the quantity of water appropriated is that for diesel fuel, lubrication oil and spare parts, particularly for the pump. The co-operators did not know what this amounted to for pumping their water from the stream. The calculation was made for my benefit and the figure for this marginal cost is £0.22 (US$0.36) per cubic metre of water supplied to the "old" storage tank at the top of the farm. The fact that the statistic was not known and is not used at Keveral suggests that the elasticity of demand for water is negligible. As a rule of thumb in the economics of irrigation, one can assume that, where a farming family or agribusiness knows neither the price nor the marginal cost of its water supply, the elasticity of demand is zero.

The expenditure of the cooperators' own labour-time in the operation and maintenance of their irrigation water supply was clearly important to them. However, irrigation also brings with it labour-time reduction in the cultivation process. This is because, for regularly irrigated crops on this farm, the distribution of organic fertilizer to the plants is carried out by attaching a bottle or bag containing the fertilizer at the head of the on-field irrigation network. As a result, the labour costs of fertilizer application are virtually eliminated.

A farm budget recalculated from Keveral Farm's own cash-flow accounts. Total costs are some £62,000, of which 86 per cent is made up by vegetable purchases, wages, taxation and capital repayments. Total income is some £82,000, of which 91 per cent is made up of vegetable re-sales, miscellaneous products and irrigated crop sales. The purchase and re-sale of vegetables bought in from other organic farms is the largest single item for both the expenditure and income columns. These vegetables plus the Farm's irrigated output are, of course, the source of the organic veggie boxes. Clearly, Keveral Farmers Lt could raise its output in value-added terms through substituting these purchased inputs by additional output of its own produce. Gross margin per acre can be calculated as total income (excluding grants) minus prime costs divided by the farm's 30 acres (12.2 ha). The figure is £810. The gross margin for the 5 acres (2.0 ha) of irrigated land could not be calculated because prime costs are not separable into irrigation/ non-irrigation outputs. We can now also calculate one of the most widely used efficiency measures in irrigation economics. We shall call this E_7, *turnover efficiency*.

$$E_7 = \frac{I}{S}$$

where I is gross sales come (turnover) received by the farm for its irrigated produce. In this case the value is £19,331/600 m^3 or £32/m^3.

The cost of water supplies is recorded as zero. The cost of water is hidden in entries such as diesel fuel and capital repayments. Returning to the prime cost of irrigation water in terms of diesel fuel, lubricants and spare parts, this equals £110/year or 0.2 per cent of the prime costs of £51,000. This reinforces the suggestion above that at Keveral Farm the elasticity of demand for pumped water is zero. As the cooperator whom we interviewed said about of the farm's prime cost outlays on water from the north stream: "...that's such a minimal cost for endless amounts of water, that we don't worry about it".

Table:2. Keveral Farmers Lt Farm Budget for the Year 1999

Costs	£	*Income*	£
Prime Costs		*Irrigated Crops*	
Vegetable purchases	28,530	Potatoes	4,867
Cooperators' wages	15,680	Cucurbits	2,740
Farm machinery and vehicles maintenance	2,050	Root vegetables	2,461
Diesel fuel and lubrication oil	1,680	Legumes	2,406
Training expenditures	720	Brassicas	1,850
Telephone	640	Aliums	1,546
Advertising	460	Tomatoes	1,425
Van hire	330	Salads	1,020
Stationery and postage	280	Soft fruit/luxury vegetables	812
Other costs	140	Aubergines and peppers	204
Water for irrigation and other uses	0		
Total prime costs	50,510	Total irrigated crops	19,331
Overhead costs		*Other farm income*	
Taxation	4,880	Vegetable resales	28,530
Capital repayments	4,100	Miscellaneous products	26,920
Bank charges	520	European Union grant	7,320
Loan interest	50		
Business insurance	930		
Soil Association and planning fees	770		
Accountancy fees	400		
Total overhead costs	11,650	Total other farm income	62,770
Total costs	62,160	Total income	82,101

Finally, we come to Keveral's on-field distribution of irrigation water. The layout and costs for the farm's twin-span, polythene greenhouse will illustrate this. A 50-mm medium-density polyethylene (MDPE) pipe runs down alongside the fields and parallel lengthways with the greenhouse. At right angles off this is a 32-mm MDPE pipe that traverses externally the whole width of the twin-span greenhouse past its two entrance flaps.

A 32-mm pipe, one for each entrance, then makes the short run into the greenhouse where a full-flow, quarter-turn ball valve is located. Quickly in sequence come the diluter bottle containing Keveral's own "wild magic" liquid fertilizer and then the timer for irrigation scheduling. This leads to a 20-mm low-density polyethylene (LDPE) pipe set at right-angles that traverses internally the width of the greenhouse, parallel to the external 32-mm distributor.

Off this main internal distributor are set at right-angles twenty-four short 16-mm pipes linked through a valve either to 16-mm T-tape or 16-mm porous pipes, each of which runs the full length of the greenhouse. It is the T-tape and the porous piping from which irrigation water is discharged. T-tape is a plastic, thin-walled, 0.55-bar dripline discharging at 1 litre/hour/metre. Porous pipe is made from recycled car tyres and is rated at 0.5 bars, discharging at 3 litres/hour/metre.

In May 2000 the capital costs of the irrigation system for the twin-span greenhouse totalled £400 inclusive of value added tax (VAT). This covered the external 32-mm piping, the internal 20-mm piping, the T-tape (to be replaced every 3 years), the porous piping, the gate valve at each entrance of the greenhouse, two diluters with their connectors, two timers with their connectors and twentyfour dripline valves. The greenhouse is 22 m by 17 m in size with a floor area of 374 m^2, giving a capital cost for the irrigation technology of £1.07/m^2.

Note that we have *not* included the capital cost of the greenhouses themselves. As has already been shown in this case study, the greenhouses are the basis of the cultivation system-other than the open fields–which necessitates irrigation but is not itself part of the off- or on-field supply infrastructure for crop water requirements. The irrigation system does not need the greenhouses, but the greenhouses are useless without the irrigation system.

THE COST OF WATER TO THE FARMER

The economic philosophy of this book from start to end is to construct economic theory in order to apply it to real life and to carry through this building process in a theoempiric manner. The Brazilian case study in particular showed that, to understand farmers' values, beliefs, motives and behaviour, such distinctions are vital. The enormous psychological distance between the Curu Valley's subsistence farmers and the three big agribusinesses is crystal-clear. Similarly, the way in which cognition of crop water requirements shifts between the texts and tabulations of the agronomic scientist and the tactile knowledge of the organic farmer becomes evident at Keveral Farm. The two case studies have confirmed how critically dependent is theory in this respect on the contingent circumstances of time and place. Price theory requires prices or own-costs or indirect fees *to exist* if it is to have any chance of "biting" on reality. It is true that own-supply costs are present in the Brazilian and UK case studies. Despite this it becomes apparent that, where the outlay on water is very low as a proportion of prime costs, water is treated by the farmer as an agronomic input, not as an economic commodity.

This at least is consistent with theory. We saw that at a price of 5 in the case of the linear and arc demand functions is estimated to be as little as "0.05. Neither in Cornwall nor in the Curu Valley did farmers know in volumetric terms how much irrigation water they are using. A subsidiary matter, touched on briefly at various points, is crop choice. When the twin-motor pump of elasticity is running–costs to the farmer that vary with water volume used and costs that play a visible part in prime costs–then the variation between crops in their water requirements may be important in the selected cropping pattern.

This is also true when the supply of water is limited. In both cases E_6, gross margin efficiency, can play a useful role in farmer decision–making. Limitations on the supply of water, such as in arid and semi-arid countries, can also be significant in the farmer's choice of technique. Greenhouses with drip/trickle field distribution are the basis of Keveral Farm's output.

Chapter 3

Integrated Water Resources Management

Integrated water resources management is the practice of making decisions and taking actions while considering multiple viewpoints of how water should be managed. These decisions and actions relate to situations such as river basin planning, organization of task forces, planning of new capital facilities, controlling reservoir releases, regulating floodplains and developing new laws and regulations. The need for multiple viewpoints is caused by competition for water and by complex institutional constraints. The decision-making process is often lengthy and involves many participants.

COMPONENTS AND VIEWPOINTS

Integrated water resources management begins with the term "water resources management" itself, which uses structural measures and nonstructural measures to control natural and human-made water resources systems for beneficial uses. Water-control facilities and environmental elements work together in water resources systems to achieve water management purposes.

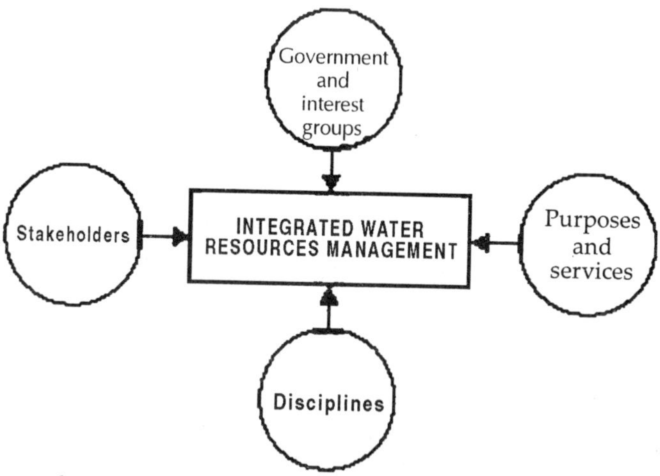

Integrated water resources management considers viewpoints of human groups, factors of the human environment and aspects of natural water systems. Structural components used in human-made systems control water flow and quality and include conveyance systems (channels, canals and pipes), diversion structures, dams and storage facilities, treatment plants, pumping stations and hydroelectric plants, wells and appurtenances.

Elements of natural water resources systems include the atmosphere, watersheds (drainage basins), stream channels, wetlands, floodplains, aquifers, lakes, estuaries, seas and the ocean. Examples of nonstructural measures, which do not require constructed facilities, are pricing schedules, zoning, incentives, public relations, regulatory programmes and insurance.

Multiple Purposes

Integrated water resources management considers the viewpoints of water management agencies with specific purposes, governmental and stakeholder groups, geographic regions and disciplines of knowledge. These viewpoints have been described in a variety of ways. For example, Mitchell wrote that integrated water management considers three aspects: dimensions of water (surface water and groundwater and quantity and quality); interactions with land and environment; and interrelationships with social and economic development. The "multiple purposes" and "multiple means" of water management, he predicted that, in integration would create some confusion because it defies neat administrative organization.

In general, water agencies deal with water supply, wastewater and water quality services, stormwater and flood control, hydropower, navigation, recreation and water for the environment, fish and wildlife. As the practice of water resources management evolved, the term "multipurpose" (or "multiobjective") water resources development (or management) came to refer to projects with more than one purpose. Later, the term "comprehensive" water planning and management came into use to describe management practice that considers different viewpoints.

Challenges to Water Management Integration

The term "functional integration" means to join purposes of water management such as to manage water supply and wastewater within a single unit.

Protecting aquatic habitat for natural and ecological systems while managing for flood control is another example. Still another term is "conjunctive use", which usually refers to the joint management of surface water and groundwater.

Governmental and Interest Groups

Accommodating the views of governments and special interest groups is a challenge in integration because they have different perspectives. Intergovernmental relationships between government agencies at the same level include regional, state-to-state and interagency issues. Relationships between different levels of government include, for example, state–federal and local–state interactions.

Special interest groups range from those favouring development of resources to those favoring preservation. In many cases, conflicts arise between the same types of interest groups, as, for example, between fly fishers and rafters on a stream.

Geographic Regions

The views of stakeholders in different locations must be balanced, introducing a geographic dimension of integration. Examples include issues between upstream and downstream stakeholders, issues among stakeholders in the same region and views of stakeholders in a basin of origin versus those in a receiving basin. Another aspect of geographic integration is the scale of water-accounting units, such as small watershed, major river basin, region, or state, even up to global scale.

Interdisciplinary Perspectives

The complexity of integrated water resources management requires knowledge and wisdom from different areas of knowledge, or disciplines. Blending knowledge from engineering, law, finance, economics, politics, history, sociology, psychology, life science, mathematics and other fields can bring valuable knowledge about the possibilities and consequences of decisions and actions. For example, engineering knowledge might focus on physical infrastructure systems, whereas sociology or psychology might focus on human impacts.

Coordination and Cooperation

Coordination is an important tool of integration because the arena of water management sometimes involves conflicting objectives. Coordinating mechanisms can be formal, such as intergovernmental agreements, or informal, such as local watershed groups meeting voluntarily. Cooperation is also a key element in integration, whether by formal or by informal means. Cooperation can be any form of working together to manage water, such as in cooperative water management actions on a regional scale, often known as "regionalization". Examples of regionalization include a regional management authority, consolidation of systems, a central system acting as water wholesaler, joint financing of facilities, coordination of service areas, interconnections for emergencies and sharing of personnel, equipment, or services.

Total Water Management

Integrated water resources management can take different forms and is examined best in specific situations. In the water-supply field, the term "integrated resource planning" has come into use to express concepts of integration in supply development. Perhaps the most comprehensive concept for water supply is "Total Water Management".

According to a 1996 report of the American Water Works Research Foundation, Total Water Management is the exercise of stewardship of water resources for the greatest good of society and the environment. A basic principle of Total Water Management is that the supply is renewable, but limited and should be managed on a sustainable-use basis. Taking into consideration local and regional variations, Total Water Management:

- Encourages planning and management on a natural water systems basis through a dynamic process that adapts to changing conditions.
- Balances competing uses of water through efficient allocation that addresses social values, cost effectiveness and environmental benefits and costs.
- Requires the participation of all units of government and stakeholders in decision-making through a process of coordination and conflict resolution.
- Promotes water conservation, reuse, source protection and supply development to enhance water quality and quantity;
- Fosters public health, safety and community goodwill.

This definition focuses on the broad aspects of water supply. Examples can be given for other situations, including water-quality management planning, water allocation and flood control.

BALANCING DIVERSE INTERESTS

Water is a natural resource critical to the environment. Water also is an economic resource critical to society. Unfortunately, people who champion water as an economic resource essential to society commonly see themselves in direct opposition to those who champion water as a critical component of the environment that must be conserved and protected. This creates a difficult situation for those who must manage the available water resources to best satisfy all of the diverse interests that depend on water. It is, of course, impossible to successfully manage water resources solely from either of the two extremes. If public policy exclusively protected ecological interests at the expense of economic interests, then it could threaten society's ability to meet the basic need all people have for water. From an ecological perspective, this could weaken government and diminish the ability to protect the environment, whereas from an economic standpoint, jobs and essential services such as housing and communications could be lost. Likewise, if water management policies focus entirely on the economy, then the environment often is harmed.

A diminished environment affects the public health, the quality of life and ultimately the ability to survive on planet Earth. Clearly, society is as dependent on a healthy environment as it is on a healthy economy.

LEGISLATION

In many Western Hemisphere countries, water resources are managed through legislation and regulations. In the United States, elected officials in Congress and the state legislatures set the policies that prioritize where water can or cannot be used and how much water must remain to function in the natural environment. These laws are then implemented through regulatory agencies at the federal, state and local levels of government. One example of this legislation is the federal Clean Water Act, which requires that the quality of surface-water bodies, such as rivers and lakes, be kept at a level that will assure the maintenance of both environmental and economic functions.

Public Opinion

The effectiveness of these policies to meet both environmental and economic interests is dependent, in part, on how well the laws are written and how effectively they are enforced. However, public opinion strongly affects the type of policies that get enacted. For example, in the 1930s, the United States was emerging from a severe economic depression, so the growth and protection of the economy were paramount in the minds of the public and policymakers alike. As a result, the national policy from roughly 1930 to 1970 was to use the water of major rivers to provide jobs and to allow cities to grow in order to expand the economy. This was often accomplished through the federal government's dam-building projects for hydroelectric power and water supply, particularly for irrigation.

Shifting Sentiment

The dam-building policy was beneficial to the economy, but by 1970 national policy and public sentiment had shifted towards the environmental health of rivers and lakes, many of which had been harmed by the dam projects. One outgrowth of this environmental pendulum shift was the Endangered Species Act (ESA) passed by the US Congress in 1973.

The ESA is an example of legislation that fundamentally changed water policy without directly addressing water issues. On its face, the law would not seem to be about water policy. When applied to fish species, however, it places requirements on water resource management to meet certain stream flow levels in order to protect a given species. Though this is clearly a good policy, it often requires extraordinary efforts in the regional management of water to ensure that the water demands of the society are met without further endangering the listed species. These efforts can be very costly. A balance must be found whereby the environment is protected to maintain the function of the rivers and lakes while still meeting the water demands of the community.

WATER MANAGEMENT

Even without the demands of environmental laws, minimizing impacts on streams and rivers while maintaining water supplies for cities and industry is a complex task because of natural constraints on water availability. Areas of the United States like the arid Southwest have limited amounts of available water such that any additional use has the potential to significantly harm natural systems. However, water resource management is difficult even in areas with substantial annual rainfall. Here the problem is often one of timing because water may be abundant only in the winter months when water demand is typically low.

Conversely, satisfying water demands for irrigation, meeting minimum-streamflow regulations and providing water to cities is typically far greater in the summer months. All of this complicates the job of the water resource manager. As with any problem where the timing of supply is not well matched with the timing of demand, the best solution is effective storage. For example, if the annual floods of rivers like the Mississippi River could be captured and stored for use later in the year, drought would be less of a problem. In this way, sufficient water is often available to meet both ecological and economic needs provided the problem is considered regionally and over a long enough period of time.

The world is now in an era where neither the economy nor the environment can absorb much negative impact from the mismanagement of water resources. Regional watershed management may help shape policy to store water from the high-flow periods of watersheds and use that water to maintain flows for fish and wildlife requirements and meet the demands of the community during low-flow periods. Everyone should continue to strive for balanced policies that allow for the use of water without inordinate harm to either the environment or the economy. The need to balance the various and diverse interests that depend on water is more essential and more complex than ever.

PLANNING AND MANAGEMENT OF WATER RESOURCES

Most public utilities engage in some form of planning, although the extent and scope of planning vary greatly. Utility planning can be characterized by four general approaches: traditional supply planning, least-cost utility planning, integrated resource planning and total water management. Each iteration retains the tried-and-true methods of the preceding approach while expanding the scope of the planning horizon to include new issues and potential solutions.

Traditional Supply Planning

Traditional planning for water utilities is not that different from traditional planning by electricity utilities, which can be characterized by its focus on utility ownership and control of all production resources (including centralization of power plants), its reliance on system and financial planning processes internal to the utility and its emphasis on the goals of minimizing electricity prices and maintaining a high level of system reliability. Risk avoidance also appears to be a paramount consideration in traditional utility planning. In the case of water, ownership and control of resources is more constrained (for example, by limits to groundwater withdrawals). However, the emphasis on utility ownership of the water delivery infrastructure (plants, pipes and so on) prevails much like the case of electricity distribution.

Shortcomings

Three principal concerns about the traditional approach have been advanced. *First,* forecast demand is taken as a given and virtually no attempt is made to integrate supply management and demand management options. *Second,* the public-at-large, outside experts and government regulators generally have little or no involvement in traditional utility planning. Demand analysis and the assessment of supply alternatives takes place within the utility (or a single planning unit within the utility); only the final product is made available for review or regulatory approval. Often, major investment decisions are made with little or no oversight. *Third,* traditional planning also tends to be confined to individual utilities in virtual isolation. The decision-making process excludes parties who not only have a vested interest but also unique insights about resource options.

In sum, there is a tendency for traditional planning to be narrowly focused and exclusionary. Once a utility has committed itself to a plan, little room is left for altering the chosen course.

Supply-driven Focus

Like other types of public utilities, the prevailing planning processes undertaken by water utilities have been internally driven and dominated by supply considerations. The result has been an emphasis on maintaining reliable water supplies and, accordingly, the engineering of facilities for source development, treatment and storage and transmission and distribution of water.

Water-supply planning generally takes the form of forecasting future demand and developing and analyzing supply options to meet the projected demand level, plus a comfortable margin. The result is a disaggregated planning approach focusing only on new supply alternatives, while initiatives in the areas of demand management and conservation are consigned to separate programmes. Moreover, conservation effects often are not reflected in revised demand and revenue forecasts.

Delivery, Not Demand

Engineering considerations have always been central to water utility planning, at times to the exclusion of other perspectives. Engineering does take into account the benefits and costs of supply projects. However, the emphasis on supply options can result in additions to capacity that outpace growth in demand, which is a problem familiar to the electricity industry. Water's natural abundance in many areas may explain why the central water supply issue is that of engineering water delivery (getting water to where it is needed) rather than managing water demand.

Historically, it was easy to presume that there always would be enough water to go around and that plentiful amounts would generally be so inexpensive that investments in demand management would not be cost-effective. The utility corporate culture, which understandably emphasizes selling more (not less) product, might also favour the supply side. At the very least, utilities traditionally have had little incentive to promote conservation.

Least-Cost Planning

The shortcomings of traditional planning, along with significant external economic and political forces, gave rise to the current interest in least-cost planning, or integrated resource planning (IRP). Least-cost planning emerged in the context of the energy industries during the 1980s as a response to rising costs, a poor record of forecasting and growing concerns about environmental externalities. According to its proponents, planning allows regulators to be more proactive, that is, to actively affect utility decisions rather than simply react to them later on. Least-cost planning emphasizes a balanced consideration of supply management and demand management options in identifying feasible least-cost alternatives for meeting future water needs. Compared with traditional planning, least-cost planning recognizes that water demand is malleable and that forecast demand does not have to simply be taken as a given in the planning process. Different definitions emphasize the minimization of rates, customer bills, utility revenue requirements and production (both capacity and operating) costs. The variety of available perspectives leaves open the question of whose costs are supposed to be least: existing customers, future customers, utility shareholders, or society at large.

Another unfortunate implication of the concept of least cost is that it suggests a single, optimal solution. Regulators and utility managers do not necessarily have identical views about the meaning of least-cost planning. According to one survey, "regulators tend to view least-cost planning with an emphasis on conservation, whereas utilities tend to regard least-cost planning as an integrated supply-and-demand analysis".

A Broader Context

Least-cost planning has come to be understood as the comprehensive evaluation of all supply and demand alternatives with the end result, in an attempt to minimize costs, of creating a flexible plan allowing for uncertainty and a changing economic environment. Cost minimization, resource diversity, risk management and flexibility are the hallmarks of least-cost planning.

In practical terms, least-cost planning can be characterized by the consideration of a diverse set of resource options (including conservation, load management, pricing and purchases from other producers); coordination among several departments within the utility; and the use of multiple resource selection goals, including those that address prices, costs, revenue requirements, utility financial condition, risk reduction, technological diversity, environmental quality and economic development. Least-cost planning in no way abandons sound engineering practices, but it does place them in a wider context. It follows that least-cost planning can facilitate regulatory review of supply management and demand management practices and regulatory approval of capacity expansion projects prior to construction.

Least-cost planning also can be used to educate policymakers about the critical issues in utility supply decisions and induce utilities to aggressively engage in long-term strategic planning. Least-cost planning can focus on a particular utility or take the form of a statewide assessment of all utilities in a given utility sector, the latter of which is typically performed by a state resource management agency.

Demand Versus Supply

In contrast to traditional supply planning, leastcost planning gives much weight to the distinction between demand management and supply management activities, both of which can be used to achieve reliability goals. Least-cost planning recognizes that demand is malleable rather than simply a fixed input to planning models. The demand side involves any strategy to eliminate or defer the need for an investment in new capacity by the utility, including load management, conservation and pricing strategies. The supply side involves determining the most efficient method of meeting growing demand, including investments in new capacity.

All demand management activities that decrease the demand for utility services tend to affect supply management since existing system capacity is released for other customers and other uses. That is, the freed or redirected utility capacity can be compared to that provided by more traditional means. Thus emerged the concept of "megawatts", meaning electricity "produced" through conservation and efficient use of existing electricity supply resources. A related idea is the trend to establish energy service companies, as compared with traditional electric utilities, who could market efficiency (or demand management) as well as electrical power.

Complexity, Risk and Cost

Least-cost utility and planning is complicated by the lack of familiarly with demand management, barriers to coordination with nearby utilities, the use of broad definitions of costs (such as those associated with externalities) and the inclusion of goals (such as resource protection) not directly attributable to the utility. Traditional planners tend to view least-cost planning as more risky, as least in terms of supply reliability. Advocates of least-cost planning would counter that planning actually reduces some forms of risk, such as the revenue recovery risk associated with excess capacity.

While highly pertinent to the debate, these disagreements can be overcome. One issue that merits especially careful attention, however, is that of the cost of planning itself. Certainly, the benefits of planning should outweigh the costs. This calculation itself poses a rational decision-making problem. The benefits of increased awareness and understanding and reduced ignorance and uncertainty can be substantial but are not easily quantified.

Integrated Resource Planning

Integrated resource planning is a somewhat more encompassing term than least-cost utility planning, although the two are consistent and can be used interchangeably for many analytical purposes. In fact, the term "least-cost integrated resource planning" sometimes is used. The concept of integrated planning evolved in part to address the potential misconceptions and complexities arising from use of the term "least-cost" as well as any unjustified bias against supply-side solutions.

Comprehensive and Participatory

Integrated resource planning encompasses the concept of least-cost planning, which emphasizes balancing supply and demand management considerations and identifying feasible planning alternatives that meet the test of least cost without unduly sacrificing other policy goals. Integrated resource planning is a more comprehensive evaluation system that goes further to emphasize the construction of various planning scenarios in which key variables and assumptions can be altered. These scenarios can be used to help utilities incorporate uncertainties, environmental externalities and community needs into decision-making.

Integrated resource planning also emphasizes the importance of establishing a more open and participatory decision-making process and coordinating the many water institutions that govern water resources. Thus, IRP encourages the development of new institutional roles in addition to new analytical tools. It also promotes consensus building and alternative dispute resolution over conflict and litigation. Importantly, planning does not preclude the development and use of markets or market-like mechanisms (such as competitive bidding), which in fact may be essential for the purpose of identifying least-cost opportunities.

Integrative Assessments

Like least-cost planning, IRP explicitly recognizes that demand management can be a cost-effective and viable resource option. In a somewhat broadened sense, IRP recognizes that demand management can help achieve multiple policy goals (such as cost control and pollution prevention). Advocates of IRP have long argued that better planning methods are needed to account for environmental and social externalities associated with expanding utility capacity. More recently, analysts have recognized that IRP can help utilities deal with uncertainties and risks as well.

The generalized concept of IRP also can be used to address short- and long-term community needs that span from environmental protection to economic development. Prescreening resource options and the construction of alternative planning scenarios can be used to evaluate the implications of a given resource mix on the utility, the environment and the community. In addition to integrating resource options, IRP also seeks integration along temporal (short- and long-term) and spatial (local and regional) dimensions.

Internal and External Linkages

Integrated resource planning clearly entails new roles and responsibilities for water-supply utilities. Integration means that environmental, engineering, public health, financial, rate-making, social and economic considerations all feed into the planning process. Planning data and information are linked internally to the other management activities of the water utility (physical facilities management, financial management, environmental management, research and development, economic development and public involvement).

Integrated planning also links water utility planning with external planning processes (planning by other water, wastewater and energy utilities; local and regional planning; river basin planning; and statewide, interstate and federal water planning and policy). Some of these relationships are formal (as in permit processes involving state water resource or drinking-water regulators), while others are less so (as in the use of regional water planning data by the utility to develop forecasts).

For the water sector, a particularly important issue is the relationship of water utility planning to the activities of various government agencies whose policies may constrain utility planning choices. Clearly, better coordination mechanisms could be established among the three or more state agencies that regulate water systems from the public health, natural resource and economic perspectives. If IRP does nothing else but facilitate coordination and improvement in the institutional structures governing water resources, it will be worth the effort.

Total Water Management

In the 1990s, total water management emerged as a potentially salient concept for water and wastewater utilities. Total water management reflects the philosophy that water resources should be managed for the greatest good of people and the environment with opportunities for participation in water policy by all segments of society.

While not strictly a planning model, total water management seems to encompass the basic principles of IRP. It also has considerable symbolic connotations. According to a white paper by the American Water Works Association, "Total water management recognizes the paradigm shift from considering water available in unlimited quantities to understanding water supply as a limited resource." Total water management seeks to inspire the water industry to embrace such ideas as sustainability, stewardship, unified water resource policies, watershed and ecosystem management, water conservation and the importance of public and political support for water management decisions. Total water management recognizes that water resources are a part of numerous complex systems, both natural and social. Advocates of integrates resources planning (encompassing, for example, water, energy and land-use planning) make a similar point. These perspectives present numerous intellectual, analytical and evaluative challenges. The term "total water management" and its manifestations are more normative and prescriptive than the seemingly value-neutral realm of rational analysis. The explicit consideration of values in planning can be uncomfortable, particularly for members of the many science-based disciplines involved in water resource issues. But in making policy choices, trade-offs among competing values are inevitable. A paradigm that allows for a dialogue about how to make these trade-offs should be welcomed by the water industry.

Planning in Practice

The practice of planning requires attention to the potential barriers to success. Three key areas of concern are: (1) access to analytical tools and adequate information, (2) the level of commitment of utilities and regulators to considering and pursuing new options and (3) the consistency of approaches and methods within the real-world context of existing utility and regulatory practices. If these issues are relevant to least-cost planning for energy utilities, they are as much or more applicable to the case of water.

Information Base

For water systems, information resources vary substantially and the need to develop data processing and analytical capabilities is clear. Attention should be paid to the design and implementation of planning strategies that minimize the effects of inadequate information.

Participants

The second barrier involves the attitudes and dispositions of those involved in resource planning. In part, success in IRP will depend on whether the prevailing corporate culture accepts conservation and related concepts as legitimate utility goals.

Realistic Approaches

Finally, successful IRP requires good practice as well as good theory. Water utility managers need practical methods for dealing with new forms of uncertainty, such as the revenue uncertainty associated with conservation.

Chapter 4

Impact of Water Pollution on Agriculture

Second only to availability of drinking water, access to food supply is the greatest priority. Hence, agriculture is a dominant component of the global economy. While mechanization of farming in many countries has resulted in a dramatic fall in the proportion of population working in agriculture, the pressure to produce enough food has had a worldwide impact on agricultural practices. In many countries, this pressure has resulted in expansion into marginal lands and is usually associated with subsistence farming.

In other countries, food requirements have required expansion of irrigation and steadily increasing use of fertilizers and pesticides to achieve and sustain higher yields. Sustainable agriculture is one of the greatest challenges. Sustainability implies that agriculture not only secure a sustained food supply, but that its environmental, socio-economic and human health impacts are recognized and accounted for within national development plans. It is well known that agriculture is the single largest user of freshwater resources, using a global average of 70% of all surface water supplies. Except for water lost through evapotranspiration, agricultural water is recycled back to surface water and/or groundwater. However, agriculture is both cause and victim of water pollution. It is a cause through its discharge of pollutants and sediment to surface and/or groundwater, through net loss of soil by poor agricultural practices and through salinization and waterlogging of irrigated land.

It is a victim through use of wastewater and polluted surface and groundwater which contaminate crops and transmit disease to consumers and farm workers. Agriculture exists within a symbiosis of land and water and, as FAO (1990) makes quite clear, "*... appropriate steps must be taken to ensure that agricultural activities do not adversely affect water quality so that subsequent uses of water for different purposes are not impaired.*"

Sagardoy summarized the action items for agriculture in the field of water quality as:

- Establishment and operation of cost-effective water quality monitoring systems for agricultural water uses.

- Prevention of adverse effects of agricultural activities on water quality for other social and economic activities and on wetlands, *inter alia* through optimal use of on-farm inputs and the minimization of the use of external inputs in agricultural activities.
- Establishment of biological, physical and chemical water quality criteria for agricultural water users and for marine and riverine ecosystems.
- Prevention of soil runoff and sedimentation.
- Proper disposal of sewage from human settlements and of manure produced by intensive livestock breeding.
- Minimization of adverse effects from agricultural chemicals by use of integrated pest management.
- Education of communities about the pollution impacts of the use of fertilizers and chemicals on water quality and food safety.

This chapter deals specifically with the role of agriculture in the field of freshwater quality. Categories of non-point source impacts–specifically sediment, pesticides, nutrients and pathogens–are identified together with their ecological, public health and, as appropriate, legal consequences. Recommendations are made on evaluation techniques and control measures. Much of the scientific literature on agricultural impacts on surface and groundwater quality is from developed countries, reflecting broad scientific concern and, in some cases, regulatory attention since the 1970s. The scientific findings and management principles are, however, generally applicable worldwide. This discussion does not deal with water quality impacts caused by food processing industries insofar as these are considered to be point sources and are usually subject to control through effluent regulation and enforcement.

WATER QUALITY AS A GLOBAL ISSUE

Agriculture, as the single largest user of freshwater on a global basis and as a major cause of degradation of surface and groundwater resources through erosion and chemical runoff, has cause to be concerned about the global implications of water quality. The associated agrofood-processing industry is also a significant source of organic pollution in most countries. Aquaculture is now recognised as a major problem in freshwater, estuarine and coastal environments, leading to eutrophication and ecosystem damage. The principal environmental and public health dimensions of the global freshwater quality problem are highlighted below:

- Five million people die annually from water-borne diseases.
- Ecosystem dysfunction and loss of biodiversity.

- Contamination of marine ecosystems from land-based activities.
- Contamination of groundwater resources.
- Global contamination by persistent organic pollutants.

Experts predict that, because pollution can no longer be remedied by dilution (i.e., the flow regime is fully utilized) in many countries, freshwater quality will become the principal limitation for sustainable development in these countries early in the present century. This "crisis" is predicted to have the following global dimensions:

- Decline in sustainable food resources (e.g., freshwater and coastal fisheries) due to pollution.
- Cumulative effect of poor water resource management decisions because of inadequate water quality data in many countries.
- Many countries can no longer manage pollution by dilution, leading to higher levels of aquatic pollution.
- Escalating cost of remediation and potential loss of "creditworthiness".

The real and potential loss of development opportunity because of diversion of funds for remediation of water pollution has been noted by many countries.

At the 1994 Expert Meeting on Water Quantity and Quality Management convened by the Economic and Social Commission for Asia and the Pacific (ESCAP), Asian representatives approved a declaration which called for national and international action to assess loss of economic opportunity due to water pollution and to determine the potential economic impacts of the "looming water crisis". Interestingly, the concern of the delegates to the ESCAP meeting was to demonstrate the economic rather than simply the environmental impacts of water pollution on sustainable development.

Creditworthiness is of concern insofar as lending institutions now look at the cost of remediation relative to the economic gains. There is concern that if the cost of remediation exceeds economic benefits, development projects may no longer be creditworthy. Sustainable agriculture will, inevitably, be required to factor into its water resource planning the larger issues of sustainable economic development across economic sectors. This comprehensive approach to management of water resources has been highlighted in the World Bank's policy on water resource development.

Older chlorinated agricultural pesticides have been implicated in a variety of human health issues and as causing significant and widespread ecosystem dysfunction through their toxic effects on organisms. Generally banned in the developed countries, there is now a concerted international effort to ban these worldwide as part of a protocol for Persistent Organic Pollutants (POPs).

NON-POINT SOURCE POLLUTION

Definition: Non-point source water pollution, once known as "Diffuse" source pollution, arises from a broad group of human activities for which the pollutants have no obvious point of entry into receiving watercourses. In contrast, point source pollution represents those activities where wastewater is routed directly into receiving water bodies by, for example, discharge pipes, where they can be easily measured and controlled. Obviously, non-point source pollution is much more difficult to identify, measure and control than point sources.

The term "Diffuse" source should be avoided as it has legal connotation in the United States that can now include certain types of point sources.

In the United States, the Environmental Protection Agency (US-EPA) has an extensive permitting system for point discharge of pollutants in water courses. Therefore, in that country, non-point sources are defined as any source which is not covered by the legal definition of "point source" Clean Water Act of 1987:

The term 'point source' means any discernible, confined and discrete conveyance, including but not limited to any pipe, ditch, channel, tunnel, conduit, well, discrete fissure, container, rolling stock, concentrated animal feeding operation, or vessel or other floating craft, form which pollutants are or may be discharged.

This term does not include agricultural storm water discharges and return flows from irrigated agriculture." The reference to "agricultural storm water discharges" is taken to mean that pollutant runoff from agriculture occurs primarily during storm flow conditions.

However, even in the United States, the distinction between point and non-point sources can be unclear and, as Novotny and Olem (1994) point out, these terms tend to have assumed legal rather than technical meanings.

Conventionally, in most countries, all types of agricultural practices and land use, including animal feeding operations (feed lots), are treated as a non-point sources. The main characteristics of non-point sources are that they respond to hydrological conditions, are not easily measured or controlled directly (and therefore are difficult to regulate) and focus on land and related management practices.

Control of point sources in those countries having effective control programmes is carried out by effluent treatment according to regulations, usually under a system of discharge permits. In comparison, control of non-point sources, especially in agriculture, has been by education, promotion of appropriate management practices and modification of land use.

Prevention and Modification of Land-use Practices

Agriculture is only one of a variety of causes of non-point sources of pollution, however, it is generally regarded as the largest contributor of pollutants of all the categories.

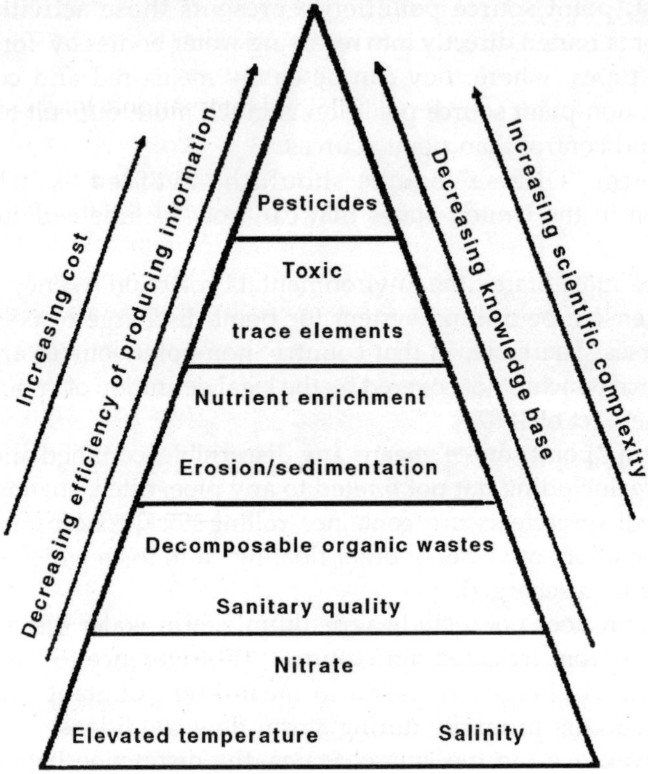

Fig:1. Hierarchical Complexity of Agriculturally-related Water Quality Problems

Scope of the Problem

Non-point source pollutants, irrespective of source, are transported overland and through the soil by rainwater and melting snow. These pollutants ultimately find their way into groundwater, wetlands, rivers and lakes and, finally, to oceans in the form of sediment and chemical loads carried by rivers. The ecological impact of these pollutants range from simple nuisance substances to severe ecological impacts involving fish, birds and mammals and on human health.

The Range and Relative Complexity of Agricultural Non-point Source Pollution

In what is undoubtedly the earliest and still most extensive study of non-point source pollution, Canada and the United States undertook a major programme of point and non-point source identification and control in the 1970s for the entire Great Lakes basin. This was precipitated by public concern over the deterioration in water quality, including the visible evidence of algal blooms and increase in aquatic weeds. Scientifically, the situation was one of hypertrophic conditions in Lake Erie and eutrophic conditions in Lake Ontario caused by excessive phosphorus entering the Lower Great Lakes from point and non-point sources.

The two countries, under the bilateral International Joint Commission, established the "Pollution from Land Use Activities Reference Groups" which served as the scientific vehicle for a ten year study of pollution sources from the entire Great Lakes basin and which culminated in major changes both to point and non-point source control. The study also resulted in an unprecedented increase in scientific understanding of the impacts of land use activities on water quality. This work, mainly done in the 1970s and early 1980s, still has great relevance to non-point source issues now of concern elsewhere in the world.

The PLUARG study, through analysis of monitoring data of rivers within the Great Lakes, from detailed studies of experimental and representative tributary catchments and from research of agricultural practices at the field and plot level, found that non-point sources in general and agriculture in particular, were a major source of pollution to the Great Lakes. By evaluation of the relative contributions of point and non-point sources to pollution loads to the Great Lakes, the PLUARG study proposed a combined programme of point source control and land use modification.

The two federal governments and riparian state and provincial governments implemented these recommendations with the result that the two lower and most impacted Great Lakes have undergone major improvements in water quality and in associated ecosystems in the past decade. A significant factor in the agricultural sector was the high degree of public participation and education.

Change in agricultural practices was, in many cases, achieved by demonstrating to farmers that there were economic gains to be realized by changing land management practices.

In most industrialized countries, the focus on water pollution control has traditionally been on point source management. In the United States, which is probably reasonably typical of other industrialized nations, the economics of further increases in point source regulation are being challenged, especially in view of the known impacts of non-point sources of which agriculture has the largest overall and pervasive impact. There is a growing opinion that, despite the billions of dollars spent on point source control measures, further point source control cannot achieve major additional benefits in water quality without significant control over non-point sources. In this context, it is relevant to note that agriculture is regarded as the main non-point source issue.

The United States is one of the few countries that systematically produces national statistics on water quality impairment by point and non-point sources. In its 1986 Report to Congress, the United States Environmental Protection Agency (US-EPA) reported that 65% of assessed river miles in the United States were impacted by non-point sources. Again, in its most recent study, the US-EPA (1994) identified agriculture as the leading cause of water quality impairment of rivers and lakes in the United States and third in importance for pollution of estuaries.

Agriculture also figures prominently in the types of pollutants. Sediment, nutrients and pesticides occupy the first four categories and are significantly associated with agriculture. While these findings indicate the major importance of agriculture in water pollution in the United States, the ranking would change in countries with less control over point sources. However, a change in ranking only indicates that point source controls are less effective, not that agricultural sources of pollution are any less polluting.

Table:1. Leading Sources of Water Quality Impairment

Rank	*Rivers*	*Lakes*	*Estuaries*
1.	Agriculture	Agriculture	Municipal point sources
2.	Municipal point sources	Urban runoff/storm sewers	Urban runoff/ storm sewers
3.	Urban runoff/storm	Hydrologic/habitat modification	Agriculture
4.	Resource extraction	Municipal point sources	Industrial point sources
5.	Industrial point sources	On-site wastewater	Resource extraction

Since the 1970s there has also been growing concern in Europe over the increases in nitrogen, phosphorus and pesticide residues in surface and groundwater. Intense cultivation and "factory" livestock operations led to the conclusion, already drawn by the French in 1980, that agriculture is a significant non-point source contributor to surface and groundwater pollution. In a recent comparison of domestic, industrial and agricultural sources of pollution from the coastal zone of Mediterranean countries, UNEP (1996) found that agriculture was the leading source of phosphorus compounds and sediment.

The European Community has responded with Directive on

"Protection of waters against pollution by nitrates from agricultural sources". The situation in France has resulted in the formation of an "Advisory Committee for the Reduction of Water Pollution by Nitrates and Phosphates of Agricultural Origin" under the authorities of the Ministry of Agriculture and the Ministry of the Environment.

Agriculture is also cited as a leading cause of groundwater pollution in the United States. In 1992, fully forty-nine of fifty states identified that nitrate was the principal groundwater contaminant, followed closely by the pesticide category. The US-EPA concluded that: *"More than 75% of the states reported that agricultural activities posed a significant threat to groundwater quality."* In an analysis of wetlands, the US-EPA (1994) reported that: *"Agriculture is the most important land use causing Wetland degradation."*

Similar data are difficult to obtain or are not systematically collected and reported in other countries, however, numerous reports and studies indicate that similar concerns are expressed in many other developed and developing countries.

AGRICULTURAL IMPACTS ON WATER QUALITY

Irrigation Impacts on Surface Water Quality

United Nations' predictions of global population increase to the year 2025 require an expansion of food production of about 40-45%. Irrigation agriculture, which currently comprises 17% of all agricultural land yet produces 36% of the world's food, will be an essential component of any strategy to increase the global food supply. Currently 75% of irrigated land is located in developing countries; by the year 2000 it is estimated that 90% will be in developing countries.

In addition to problems of waterlogging, desertification, salinization, erosion, etc., that affect irrigated areas, the problem of downstream degradation of water quality by salts, agrochemicals and toxic leachates is a serious environmental problem. "It is of relatively recent recognition that salinization of water resources is a major and widespread phenomenon of possibly even greater concern to the sustainability of irrigation than is that of the salinization of soils, per se. Indeed, only in the past few years has it become apparent that trace toxic constituents, such as Se, Mo and As in agricultural drainage waters may cause pollution problems that threaten the continuation of irrigation in some projects".

Public Health Impacts

Polluted water is a major cause of human disease, misery and death. According to the World Health Organization (WHO), as many as 4 million children die every year as a result of diarrhoea caused by water-borne infection. The bacteria most commonly found in polluted water are coliforms excreted by humans. Surface runoff and consequently non-point source pollution contributes significantly to high level of pathogens in surface water bodies. Improperly designed rural sanitary facilities also contribute to contamination of groundwater.

Agricultural pollution is both a direct and indirect cause of human health impacts. The WHO reports that nitrogen levels in groundwater have grown in many parts of the world as a result of "intensification of farming practice". This phenomenon is well known in parts of Europe. Nitrate levels have grown in some countries to the point where more than 10% of the population is exposed to nitrate levels in drinking water that are above the 10 mg/l guideline. Although WHO finds no significant links between nitrate and nitrite and human cancers, the drinking water guideline is established to prevent methaemoglobinaemia to which infants are particularly susceptible.

Although the problem is less well documented, nitrogen pollution of groundwater appears also to be a problem in developing countries. Lawrence and Kumppnarachi reported nitrate concentrations approaching 40-45 mg N/l in irrigation wells that are located close to the intensively cultivated irrigated paddy fields. The variation in NO_3-N which shows a peak in the *maha* (main) cropping season when rice growing is most intensive in Sri Lanka. Reiff, irrigated agriculture, notes that water pollution is both a cause and an effect in linkages between agriculture and human health. The following health impacts which apply, in particular, to developing countries, were noted by Reiff:

- Adverse environmental modifications result in improved breeding ground for vectors of disease (e.g., mosquitos). There is a linkage between increase in malaria in several Latin American countries and reservoir construction. Schistosomiasis (Bilharziasis), a parasitic disease affecting more than 200 million people in 70 tropical and subtropical countries, has been demonstrated to have increased dramatically in the population following reservoir construction for irrigation and hydroelectric power production. Reiff indicates that the two groups at greatest risk of infection are farm workers dedicated to the production of rice, sugar cane and vegetables and children that bathe in infested water.
- Contamination of water supplies primarily by pesticides and fertilizers. Excessive levels of many pesticides have known health effects.
- Microbiological contamination of food crops stemming from use

of water polluted by human wastes and runoff from grazing areas and stockyards. This applies both to use of polluted water for irrigation and by direct contamination of foods by washing vegetables, etc, in polluted water prior to sale. In many developing countries there is little or no treatment of municipal sewage, yet urban wastewater is increasingly being used directly or recycled from receiving waters, into irrigated agriculture. The most common diseases associated with contaminated irrigation waters are cholera, typhoid, ascariasis, amoebiasis, giardiasis and enteroinvasive *E. coli*. Crops that are most implicated with spread of these diseases are ground crops that are eaten raw such as cabbage, lettuce, strawberries, etc.

- Contamination of food crops with toxic chemicals.
- Miscellaneous related health effects, including treatment of seed by organic mercury compounds, turbidity (which inhibits the effectiveness of disinfection of water for potable use), etc.

To this list can be added factors such as the potential for hormonal disruption (endocrine disruptors) in fish, animals and humans. Hormones are produced by the body's endocrine system. Because of the critical role of hormones during early development, toxicological effects on the endocrine system often have impacts on the reproductive system. While pesticides such as DDT have been implicated, the field of endocrine disruption is in its infancy and data which support cause and effect are not yet conclusive.

It is probably safe to conclude, however, that high levels of agricultural contaminants in food and water as are found in many developing country situations have serious implications for reproduction and human health.

AGRICULTURE AND THE ARAL SEA DISASTER

The social, economic and ecological disaster that has occurred in the Aral Sea and its drainage basin since the 1960s, is the world's largest example of how poorly planned and poorly executed agricultural practices have devastated a once productive region. Although there are many other impacts on water quality in the region, improper agricultural practice is the root cause of this disaster. Virtually all agriculture is irrigated in this arid area. The Aral Sea basin includes Southern Russia, Uzbekistan, Tadjikistan and part of Kazakhstan, Kirghiztan, Turkmenistan, Afghanistan and Iran.

Population: 1976 = 23.5 million; and 1990 = 34 million

Area: 1.8×10^6 km^2

% Irrigated = 65.6% (1985)

Water Balance of the Aral Sea Basin

Perennial (average) water supply: 118.3 km^3/yr (100%)

Irrigation demand (current estimates): 113.9 km^3/yr (96.3%)

Consumptive use in irrigation is 75.2 km^3/yr (63.4% of available water supply)

Irrigation Expansion and Inflow to Aral Sea

Irrigation	:	Since 2000-3000 B.C.	
		1950s + - major expansion	
		1985 - 65.6% of total land area	
Inflow to Aral Sea	:	Historical	: 56 km^3/yr
		1966-1970	: 47 km^3/yr
		1981-1985	: 2 km^3/yr

Salinization

Magnitude and acceleration of salinization is demonstrated in Uzbekistan

	Salinized Area	% of Total Irrigated Area
1982	12 000 km^2	36.3
1985	16 430 km^2	42.8

Public Health Impacts (Over past 15 years)

Typhoid	: 29-fold increase (morbidity index up 20%)
Viral Hepatitis	: 7-fold increase
Paratyphoid	: 4-fold increase

Number of persons with hypertonia, heart disease, gastric and duodenal ulcers -up 100%

Increase in premature births - up 31%

Data on Agricultural Water Pollution in Developing Countries

Data on water pollution in developing countries are limited. Further, such data are mostly "aggregated", not distinguishing the relative proportion of "point" and "non-point" sources. In Thailand, the Ministry of Public Health reported the results of pollution monitoring of 32 rivers.

Pesticide consumption has strongly increased in all developing countries. In India, consumption increased nearly 50-fold between 1958 and 1975. Yet the Indian consumption in 1973-74 was reported to be averaging a mere 330 g/ha, compared to 1483 g in USA and 1,870 g in Europe. According to various surveys in India and Africa, 20-50% of wells contain nitrate levels greater than 50 mg/1 and in some cases as high as several hundred milligrams per litre.

In the developing countries, it is usually wells in villages or close to towns that contain the highest levels, suggesting that domestic excreta are the main source, though livestock wastes are particularly important in semi-arid areas where drinking troughs are close to wells.

Table:2. Pollution of 32 rivers in Thailand

Types of Pollution	*No. of Rivers Affected out of 32 Monitored*
Organic waste	13
Microbial waste	20
Heavy metals	8

TYPES OF DECISIONS IN AGRICULTURE FOR NON-POINT SOURCE POLLUTION CONTROL

Decisions by agriculturalists for control of agricultural non-point source pollution can be at various scales. At the field level, decisions are influenced by very local factors such as crop type and land use management techniques, including use of fertilizers and pesticides. These decisions are based on best management practices that are possible under the local circumstances and are meant to maximize economic return to the farmer while safeguarding the environment.

Local decisions are made on the basis of known relationships between farm practice and environmental degradation but do not usually involve specific assessment of farm practices within the larger context of river basin impacts from other types of sources.

Decisions regarding use of waste water, sludges, etc., for agricultural application are also made using general knowledge of known impacts and of measures to mitigate or minimize these impacts. However, the challenge for agriculturists is to mobilize the knowledge base and to make it available to farmers.

At the river basin scale, the nature of decision-making is quite different. At this scale, the typical decision-making problem for non-point source control in many developing countries. It is not possible in this publication to describe in detail the "tools" that are used to address this basin-scale management problem. Moreover, many of the tools are not yet systematized to the point where they are easily accessible to agricultural practitioners.

THE DATA PROBLEM

One area, however, that is well known, is the data problem. The water quality database that is available in many developing countries (and in some developed countries) is of little value in pollution management at the river basin scale nor is it useful for determining the impact of agriculture relative to other types of anthropogenic impacts.

A common observation amongst water quality professionals is that many water quality programmes, especially in the developing countries, collect the wrong parameters, from the wrong places, using the wrong substrates and at inappropriate sampling frequencies and produce data that are often quite unreliable. Further, the data are not assessed or evaluated and are not sufficiently connected to realistic and meaningful programme, legal or management objectives. This is not the fault of the developing countries; more often it results from inappropriate technology transfer from the developed countries and an incorrect assumption by recipients and donors that the data paradigm developed by the developed countries is appropriate in the developing countries.

Additionally, water quality monitoring programmes, worldwide, are under severe stress as governments reduce budgets, downsize and shift priorities. "Monitoring" has become a dirty word and governments are increasingly reluctant to pay for it.

Paradoxically, the need for reliable water quality information has never been greater. Fortunately, new scientific research, together with budget realities, now makes it possible to rethink and redesign data programmes that are inherently more focused, more practical, more efficient, produce more information and less data and which meet programme goals in measurable economic terms.

This publication is not the place to deal substantively with new monitoring (data collection) techniques; however, it is sufficient to note here that monitoring technology has changed dramatically in the past decade, to the point where significant economic and information gains can be achieved in most monitoring programmes. Significant for agricultural programmes is that water quality data are rarely collected by ministries of agriculture. Nevertheless, sustainable agriculture within the framework of comprehensive basin management will require relevant and reliable data upon which to make management decisions. This will necessitate intervention by agriculturalists in existing water quality data programmes if relevant data are to be collected for agricultural management purposes.

POLLUTION BY SEDIMENTS

Although agriculture contributes to a wide range of water quality problems, anthropogenic erosion and sedimentation is a global issue that tends to be primarily associated with agriculture. While there are no global figures, it is probable that agriculture, in the broadest context, is responsible for much of the global sediment supply to rivers, lakes, estuaries and finally into the world's oceans.

Pollution by sediment has two major dimensions. One is the Physical Dimension–top soil loss and land degradation by gullying and sheet erosion and which leads both to excessive levels of turbidity in receiving waters and to off-site ecological and physical impacts from deposition in river and lake beds. The other is a Chemical Dimension–the silt and clay fraction (<63 mm fraction), is a primary carrier of adsorbed chemicals, especially phosphorus, chlorinated pesticides and most metals, which are transported by sediment into the aquatic system.

Erosion is also a net cost to agriculture insofar as loss of top soil represents an economic loss through loss of productive land by erosion of top soil and a loss of nutrients and organic matter that must be replaced by fertilizer at considerable cost to the farmer in order to maintain soil productivity.

The reader is referred to Roose for a detailed analysis of the social, economic and physical consequences of erosion of agricultural land and of measures that should be taken to control erosion under different types of land use, especially in developing countries. Whereas Roose is mainly concerned with the impact of erosion on agriculture, this publication is primarily concerned with agricultural erosion from the perspective of its impacts on downstream water quality.

Control of agricultural pollution usually begins, therefore, with measures to control erosion and sediment runoff, therefore,the principal mechanisms which govern erosion processes and those measures which can be taken to control erosion.

Sediment as a Physical Pollutant

Global estimates of erosion and sediment transport in major rivers of the world vary widely, reflecting the difficulty in obtaining reliable values for sediment concentration and discharge in many countries, the assumptions that are made by different researchers and the opposing effects of accelerated erosion due to human activities (deforestation, poor agricultural practices, road construction, etc.) relative to sediment storage by dam construction.

Milliman and Syvitski (1992) estimate global sediment load to oceans in the mid-20th century at 20 thousand million t/yr, of which about 30% comes from rivers of southern Asia. Significantly, they believe that almost 50% of the global total comes from erosion associated with high relief on islands of Oceania–a phenomenon which has been underestimated in previous estimates of global sediment production. While erosion on mountainous islands and in upland areas of continental rivers reflects natural topographic influences, Milliman and Syvitski suggest that human influences in Oceania and southern Asia cause disproportionately high sediment loads in these regions.

Sediment, as a physical pollutant, impacts receiving waters in the following principal ways:

- High levels of turbidity limit penetration of sunlight into the water column, thereby limiting or prohibiting growth of algae and rooted aquatic plants. In spawning rivers, gravel beds are blanketed with fine sediment which inhibits or prevents spawning of fish. In either case, the consequence is disruption of the aquatic ecosystem by destruction of habitat. Notwithstanding these undesirable effects, the hypertrophic (nutrient rich) status of many shallow lakes, especially in developing countries, would give rise to immense growth of algae and rooted plants were it not for the limiting effect of light extinction due to high turbidity. In this sense, high turbidity can be "beneficial" in highly eutrophic lakes; nevertheless, many countries recognise that this situation is undesirable for both aesthetic and economic reasons and are seeking means to reduce both turbidity and nutrient levels.

- High levels of sedimentation in rivers leads to physical disruption of the hydraulic characteristics of the channel. This can have serious impacts on navigation through reduction in depth of the channel and can lead to increased flooding because of reductions in capacity of the river channel to efficiently route water through the drainage basin. For example, calculations by the UFRGS (1991) of erosion and sediment transport in the Sao Francisco River Basin, a large drainage system in eastern Brazil, demonstrate that the central portion of the river basin is now dominated by sediment deposition. This has resulted in serious disruption of river transportation and clogs hydraulic facilities that have been built to provide irrigation water from the main river channel. The sediment largely originates from rapidly eroding sub-basins due to poor agricultural practices.

Sediment as a Chemical Pollutant

The role of sediment in chemical pollution is tied both to the particle size of sediment and to the amount of particulate organic carbon associated with the sediment. The chemically active fraction of sediment is usually cited as that portion which is smaller than 63 mm (silt + clay) fraction. For phosphorus and metals, particle size is of primary importance due to the large surface area of very small particles.Phosphorus and metals tend to be highly attracted to ionic exchange sites that are associated with clay particles and with the iron and manganese coatings that commonly occur on these small particles. Many of the persistent, bioaccumulating and toxic organic contaminants, especially chlorinated compounds including many pesticides, are strongly associated with sediment and especially with the organic carbon that is transported as part of the sediment load in rivers.

Measurement of phosphorus transport in North America and Europe indicate that as much as 90% of the total phosphorus flux in rivers can be in association with suspended sediment. The affinity for particulate matter by an organic chemical is described by its octanol-water partitioning coefficient (K_{OW}). This partitioning coefficient is well known for most organic chemicals and is the basis for predicting the environmental fate of organic chemicals. Chemicals with low values of K_{OW} are readily soluble, whereas those with high values of K_{OW} are described as "hydrophobic" and tend to be associated with particulates. Chlorinated compounds such as DDT and other chlorinated pesticides are very hydrophobic and are not, therefore, easily analysed in water samples due to the very low solubility of the chemical. For organic chemicals, the most important component of the sediment load appears to be the particulate organic carbon fraction which is transported as part of the sediment. Scientists have further refined the partitioning coefficient to describe the association with the organic carbon fraction (K_{OC}).

Another important variable is the concentration of sediment, especially the <63 mm fraction, in the water column. Even those chemicals that are highly hydrophobic will be found in trace levels in soluble form. Where the suspended load is very small (say, less than 25 mg/l), the amount of water is so large relative to the amount of sediment that the bulk of the load of the chemical may be in the soluble fraction. This becomes an important issue in the monitoring of hydrophobic chemicals.

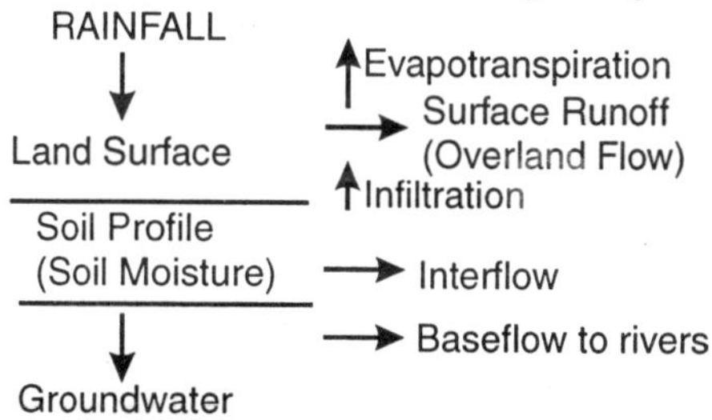

Fig:2. Schematic Diagram showing the Major Processes that Link Rainfall and Runoff

Unlike phosphorus and metals, the transport and fate of sediment-associated organic chemicals is complicated by microbial degradation that occurs during sediment transport in rivers and in deposited sediment. Nevertheless, the role of sediment in the transport and fate of agricultural chemicals, both for nutrients, metals and pesticides is well known and must be taken into account when monitoring for these chemicals and when applying models as a means of determining optimal management strategies at the field and watershed level.

For this reason, models using the "fugacity" concept (uses the partitioning characteristics of chemicals as a basis for determining the environmental compartment–air, sediment, water, biota–in which the chemical is primarily found) has proven effective in predicting the environmental pathways and fate of contaminants.

- *Conclusion:* The role of sediment as a chemical pollutant is a function of the chemical load that is carried by sediments.

Organic chemicals associated with sediment enter into the food chain in a variety of ways. Sediment is directly ingested by fish however, more commonly, fine sediment (especially the carbon fraction) is the food supply for benthic (bottom dwelling) organisms which, in turn, are the food source for high organisms. Ultimately, toxic compounds bioaccumulate in fish and other top predators. In this way, pesticides that are transported off the land as part of the runoff and erosion process, accumulate in top predators including man.

KEY PROCESSES: PRECIPITATION AND RUNOFF

The major characteristic of non-point source pollution is that the primary transfer mechanisms from land to water are driven by those hydrological processes that lead to runoff of nutrients, sediment and pesticides. This is important, not only to understand the nature of agricultural pollution, but also because modelling of hydrological processes is the primary mechanism by which agriculturalists estimate and predict agricultural runoff and aquatic impacts. Except where agricultural chemicals are dumped directly into watercourses, almost all other non-point source control techniques in agriculture involve control or modification of runoff processes through various land and animal (manure) management techniques.

In large parts of the world, precipitation is in the form of rain. However, in those areas where precipitation is in the form of snow, the science becomes more complex. Nevertheless, control measures, whether for areas subject to rain or snow can be easily summarized. Therefore, for the purpose of this publication, focus will be on the relationships between rainfall and runoff.

While the practice of hydrology can be quite theoretical, the principal concepts are easily understood.

Rainfall

The primary controlling factor is the rate (intensity) of rainfall. This controls the amount of water available at the ground surface and is closely related to measures of energy that are used in many mathematical formulations to calculate soil detachment by rain drops. Soil detachment makes soil particles available for sediment runoff.

Soil Permeability

Permeability is a physical characteristic of a soil and is a measure of the ability of the soil to pass water, under saturated conditions, through the natural voids that exist in the soil. Permeability is a function of soil texture, mineral and organic composition, etc. In contrast, "porosity" is the measure of the amount of void space in a soil; however, permeability refers to the extent to which the porosity is made up of interconnecting voids that allow water to pass through the soil. As an example, styrofoam is highly porous but impermeable, whereas a sponge is both porous and permeable.

Infiltration

Infiltration rate, the rate at which surface water passes into the soil (cm/hr), is one of the most common terms in hydrologic equations for calculating surface runoff. Infiltration is not identical to permeability; it is mainly controlled by capillary forces in the soil which, in turn, reflect the prevailing conditions of soil moisture, soil texture, degree of surface compaction, etc. Infiltration will vary between and within rainfall events, depending upon factors such as antecedent soil moisture, nature of vegetation, etc. In general, infiltration rate begins at a high value during a precipitation event and decreases to a small value when the soil has become saturated.

Surface Runoff

This is the amount of water available at the surface after all losses have been accounted for. Losses include evapotranspiration by plants, water that is stored in surface depressions caused by irregularity in the soil surface and water that infiltrates into the soil. The interaction between infiltration rate and precipitation rate mainly governs the amount of surface runoff. Intense rainstorms tend to produce much surface runoff because the rate of precipitation greatly exceeds the infiltration rate. Similarly, in areas of monsoonal rain and tropical storms, the length and intensity of precipitation frequently exceeds infiltration capacity. Destruction of protective surface vegetation and compaction of the soil, especially in tropical environments, leads to major erosional phenomena due to the amount of surface runoff. Except for nitrogen which is usually found in groundwater in agricultural areas, surface runoff is the primary contributor of agricultural chemicals, animal wastes and sediment to river channels.

Interflow

(sometimes called "throughflow") Because soil horizons have different levels of permeability not all water in the soil will move downward into the groundwater. The residual water in the soil will move along the soil horizons, parallel to the ground surface. Interflow usually emerges near the bottom of slopes and in valley bottoms. Therefore, identification of these hydrologically active zones is an important part of agricultural non-point source control measures. Interflow is the mechanism which has also been linked to soil piping, a potentially destructive characteristic in some soils by which shallow "pipes" form naturally in the soil and are enlarged by interflow to the point where they collapse causing gullies in the agricultural surface.

Groundwater

Groundwater is supplied by water which passes through the soil horizons into the parent material and/or bedrock underlying the soil. Groundwater tends to flow towards rivers channels where it emerges and supports stream flow during periods of little or no rain. This component of stream flow is called "base flow". The chemistry of base flow reflects the soil and bedrock geochemistry, plus any agrochemicals that have been leached into the groundwater.

Snowmelt

The phenomenon of snowmelt greatly complicates prediction of agricultural pollution using conventional hydrologic models. Snowmelt, by itself, is not normally a major producer of surface runoff. However, the combination of spring rain and snowmelt on frozen or thawing soils can produce serious erosional problems. Snowmelt tends to contribute greatly to agricultural non-point source pollution by carrying to adjacent streams the animal wastes, sludges and other wastes that were spread on frozen agricultural soils during the winter period. Correct management of animal wastes in regions of frozen ground has major beneficial effects on water quality.

SEDIMENT DELIVERY RATIO

The sediment delivery ratio (SDR) is commonly used in erosion and transport studies to describe the extent to which eroded soil (sediment) is stored within the basin. The SDR is defined as:

$$\text{SDR} = \frac{\text{Measured Sediment Yield}}{\text{Gross erosion in the basin}}$$

where *yield* is determined from reservoir sedimentation or from a sediment monitoring station and *gross erosion* is estimated using an estimation techniques such as the Universal Soil Loss Equation.

The SDR is always less than 1.0, indicating that soil that is eroded at the field level tends not to travel far before it is deposited. Indeed, sediment storage in rills on fields, at field margins and at the foot of slopes is large. Storage also occurs in river channels (bed and overbank deposition), in wetlands and in reservoirs and lakes. The SDR is highly variable, however the concept is one of the most important in the understanding of erosion and sedimentation processes and how these operate in time and space.

SEDIMENT ENRICHMENT RATIO

The concept of the sediment enrichment ratio (SER) is quite important in understanding the impact and economic cost of chemical loss from fields. The process of surface erosion tends to be selective towards fine particles. Consequently, the particle size characteristics of material eroded at source (at the plot level) is progressively changed towards finer particles through deposition of the coarser fraction (e.g., sand-size material). Because of the chemically enriched nature of fine particles due to the large surface area of clay-size sediment, the concentration of chemicals that are associated with sediment (phosphorus, metals, organic nitrogen, hydrophobic pesticides) increases as the impoverished sand-size fraction is lost during down-field transport resulting in an increasing proportion of the chemically enriched fine (silt-clay) fraction.

The Sediment Enrichment Ratio (SER) is defined as:

$$\text{SER} = \frac{\text{Concentration of chemical "X" in transported sediment}}{\text{Concentration of the chemical "X" in the soil}}$$

Sediment chemistry is measured at some point downslope, e.g., at the edge of a field or in adjacent streams.

The importance of the enrichment ratio lies in the fact that there is proportionally more fine-grained sediment transported than coarse-grained sediment during surface erosion. Therefore, the sediment being transported has a finer texture than the source soil material. Because of the affinity of soil nutrients for fine sediment, this proportionally larger loss of fine material means that there is net impoverishment of the soil. This usually constitutes "mining" of the natural nutrition of the soil (often referred to as "natural capital") and which may never be replaced by the addition of fertilizer. The cost to the farmer is therefore two-fold: loss of productivity due to loss of natural nutrition in the soil; and economic cost of fertilizer which is added in the attempt to compensate for this loss.

MEASUREMENT AND PREDICTION OF SEDIMENT LOSS

Prediction Models

Agriculturalists worldwide have spent much time and resources attempting to find reliable methods of predicting erosion and sediment-associated chemical runoff under different conditions of crop type, tillage practices, etc. Consequently, there is a large number of models that have been developed for the prediction of agricultural non-point source runoff of sediment, nutrients and pesticides. Many of the models permit gaming with alternative choices of land management, crop type and fertilizer and pesticide application rates. Because all models (except unit load models) require hydrometric input and many use a sediment sub-component, it is appropriate to integrate these into a single table together with their principal characteristics.

In general, there are three types of models based on input data needs.

- Simple screening models, such as the unit load approach, attempt to provide approximate answers about the likely magnitude of sediment and chemical runoff. This approach is a statistical methodology in which data on runoff of sediment, nutrients and pesticides on a unit area basis (e.g., tonnes of sediment per hectare) are collated from many studies to reflect similarities in crop type, soil and physiographic characteristics. Unlike the other types of models which are largely focused on prediction and improvement of agricultural management at the farm level, the unit load approach is mainly focused on impacts of agriculture on downstream water quality and without consideration of alternative farm management practices.
- *Simple empirical relationships:* The widely used and respected Universal Soil Loss Equation (USLE) of Wischmeier (1976) has had remarkable success at the plot level and has been incorporated into many of the complex models. The USLE is designed as a field management tool and provides aggregated information at the storm, seasonal or annual level. Wischmeier (1976) reported that the average prediction error for annual soil loss was 12%; larger errors are to be expected for single storm events. The USLE is one way to determine erosion potential for input into the denominator of the Sediment Delivery Ratio. The USLE is detailed here because of its success and because the same type of approach has been used in Africa and elsewhere.

  The USLE is calculated as:

  A = R. K. LS. C. P

where:

A = Calculated soil loss in t/ha for a given storm or period
R = Rainfall energy factor
K = Soil erodibility factor
LS = Slope-length factor

C = Cropping management factor (vegetative cover)

P = Erosion-control practice factor.

Each of the factors can be calculated or estimated using field data and from tables or nomograms for all other factors. Novotny and Olem (1994) provide an excellent commentary on this and other methods for estimating or modelling erosion. The USLE is designed for rainfall only and does not handle snowmelt or rainfall on frozen ground. The USLE requires calibration data from standard plot experiments which are widely available in North America and, more limitedly, from other parts of the world.

At the international level, the simplicity and effectiveness of the soil loss approach prompted an important series of experiments in Zimbabwe in the 1950s and 1960s with the primary objective of determining losses of nitrogen, phosphorus and organic carbon from the natural fertility of the soil. According to Stocking who exhaustively analyzed this database, it represented the "best such database in any developing or tropical country". This work led to the development of the SLEMSA model for conditions in Southern Africa. The value of this approach has further led FAO (1985) to develop an international network, "Network on Erosion-Induced Loss in Soil Productivity", with research partners in Africa, South America and Asia.

Although Wischmeier (1976) has cautioned against extending soil loss models beyond field loss studies, these models are intuitively attractive for predicting erosion over large areas. It should be noted that, due to transport losses, such erosion estimates apply only to total erosion at source and do not reflect sediment loads (or sediment yield) measured at downstream locations. Such estimating techniques would, if calibrated so that the errors are known, have useful application as a screening tool for the estimation of erosion potential under conditions of similar crop, soil and topographic factors over large areas. Internationally, there appears to be little systematic information on calibration; however, the Network on Erosion-Induced Loss in Soil Productivity may eventually provide suitable information.

The most extreme example of use of the soil loss approach for large area estimation is in Brazil where the UFRGS (1991) and Carvalho (1988) used large area maps and satellite data to estimate several of the parameters of the USLE for application at regional scales. The intent was to provide generalized estimates of regional erosion potential for the entire country. While this approach has wide margins of error, it represents a method of screening for major change in erosion potential arising from combinations of agricultural land use, climate and topography and merits further consideration, especially where plot calibration and in-river sediment monitoring data are available.

Together with the need to further develop screening level models, is the need to generate improved field data on erosion and sediment loss. Hudson presented a wide range of simple field measurement techniques that are particularly useful in developing countries.

There is a wide variety of deterministic and stochastic models that attempt to simulate the physics of the erosion process. The data requirements for calibration and verification are extremely large. While such models may have certain advantages, especially in terms of the level of detail in which one can simulate alternative farm practices, these are generally unsuitable for developing countries due to their data requirements and the observation that management judgements for farm-level decisions can almost always be made on the basis of more generalized data combined with experience and common sense.

Sediment Yield

Sediment yield, usually expressed as tonnes per unit area of the basin per year, is the amount of sediment measured at some point in the basin divided by the basin area. It always less than the total erosion due to sediment storage during transport and is highly variable because of measurement difficulty, the temporal variability of hydrological processes and changes in land management practices in the basin from one year to the next.

The values of demonstrate not only the wide range of values in different climatic and topographic situations, but also the problem of interpreting sediment data because of spatial and time scales. As noted above, the Sediment Delivery Ratio subsumes a wide range of storage processes that operate at the field, sub-basin and basin scales. Sediment yield values are also highly variable in time, with smaller basins subject to greater variability than larger basins. Consequently, published values for sediment yield (t/ha/yr) must be interpreted with great caution. The only plot data and would appear much smaller if they were based on sediment measurements taken at the sub-basin scale. The magnitude of change in sediment yield arising from changes in agricultural land.

Estimates of sediment yield have important economic consequences. In many developing countries the database with which to estimate reservoir life is very limited. According to White (1988), examples of predicted sediment yield in Asia tend to be between two and sixteen times lower than actual measured rates, with the consequence that actual reservoir life is greatly reduced. This is also a problem in northern Africa.

Partly this arises from the use of unreliable prediction techniques and the use of short-term data which usually fail to account for occasional but severe episodes of erosion (e.g., major storm events) and from increased land pressure after construction of the reservoir. White reports that the Magat Dam was constructed on the basis of a planned sediment yield of 20 t/ha/yr when, in reality, the yield was shown to be 38 t/ha/yr–cutting the useful life of the reservoir in half! In the Republic of South Africa, off-site damage due to soil erosion was estimated in 1989 to total US$ 37.6 million annually. Because agriculture is a major contributor to sediment yield it is essential that national agricultural organizations account for off-site costs.

Scale Problems

In all aspects of monitoring, modelling, or prediction of non-point source management, the scale factor is poorly understood by many practitioners. The scale factor not only pertains to the cost of collecting data for model calibration, but it is also vital to the ability to extrapolate useful management principles that will apply to larger areas. As an example, during the 1970s it was assumed by many modellers who were engaged in the non-point source management programme of the North American Great Lakes, that extrapolation of cause-effect relationships from small-area studies into larger parts of the Great Lakes basin was feasible. It was found, however, that extrapolation became quite uncertain as basin area increased, for a variety of reasons.

The problem of extrapolation has not been definitively addressed in regards to levels of uncertainty. For example, although it is known that control of agricultural erosion is essential to mitigate nutrient (especially phosphorus) loss to receiving watercourses, it is unclear over what distance phosphorus runoff from agriculture has an immediate impact. The question then is, does erosion control in far upstream areas have an immediate benefit for a receiving water that may lie hundreds of kilometers downstream? Alternatively, one may adopt the position that over longer periods of time (say, scores of years) large storms will ensure that most of the eroded sediment is, in fact, transported through the river system and, therefore, one can assume a 1:1 relationship between upstream sediment production and downstream sediment transport. Research by Meade and Trimble (1974) and others shows that over longer time periods, sediment is mobilized differentially along the river basin. Contemporary sediment mobilization in the middle reaches of piedmont rivers of the United States represented sediment that was eroded several decades earlier but which had been deposited during transport into long-term storage as valley floor deposits.

The essential point is that time and space scales must be recognized when preparing management plans for erosion control. While the near-term physical benefits of erosion control are likely to be felt quickly in receiving waters, sediment-associated contaminants that are stored in the river basin may take decades to finally be transported out of the basin, notwithstanding the resources spent on upstream erosion control.

RECOMMENDATIONS

These recommendations reflect two very different scales which, in turn, reflect two different types of issues. At the small scale are the recommendations which apply at the farm level and which reflect actions to be taken by the individual farmer after considering the economic costs of his alternatives. At the large scale (river basin scale) are those issues that tend to reflect policy and investment needs of states. This includes issues such as determining the net contribution of agriculture to river pollution relative to other types of polluting sources.

Internalizing Costs

Although the control of erosion at source in rain-fed agriculture is a major factor in improving downstream water quality and associated ecological impacts, implementation of control measures will only be successful if the farmer can determine that it is in his economic interest to undertake such measures. Therefore, the economic benefits such as maintenance of soil fertility, reduced energy consumption in minimum till situations, etc., relative to economic costs of excessive fertilizer usage and loss of productivity by "mining" of soil capital, must be clearly demonstrated. This implies that agricultural agencies must use a holistic approach to the economics of farming practices.

Screening and Estimating Tools

There is an urgent need to develop simple and robust screening models for use in developing countries to determine the potential for erosion and soil loss at source (field level). The models must contain the ability to game with crop and land management alternatives. Data requirements must be relatively small and easily accessible, require no real-time calibration and aggregate results on a seasonal or annual basis. Further calibration of the USLE (and its derivatives) and SLEMSA models are encouraged. These models must be developed from the perspective of assisting in farm-level decisions and not from the perspective of further elaboration of the physics of erosion and sediment transport.

At the large scale, screening methods are needed to develop policy options for erosion control and land use practices at the basin scale. Soil loss models require adequate calibration so that the errors inherent in large scale use, using satellite and large scale map information, are known. Use of sediment loading data from river monitoring sites is an inadequate alternative due to the large sediment losses that occur during sediment transport from field to river.

Erosion Control

There are no unique solutions to erosion control. Control measures depend very much on the economic situation of the farmer, the degree of importance placed on sediment erosion by environmental authorities, availability of capital and the state of development of the country.

These categories are used in many parts of the world, including developing countries. These techniques also have beneficial effects for conservation of nitrogen and phosphorus in the soil.

Data from Chesapeake Bay (USA) installations indicate the following ranking of costs of erosion control measures. Cost factors and rankings will vary greatly, especially in parts of the world where labour costs are less and where the economic benefit is factored into the ranking system. For example, grassed filter strips may have no economic benefit and are replaced with alternative practices such as herbaceous cover mixed with fruit trees.

Erosion in tropical areas has a unique set of problems. Marginal agricultural practices such as slash and burn on highly erodible tropical soils and tillage on steep tropical slopes lead to highly unstable situations which erode quickly during rainy seasons. Similarly, deforestation in tropical lands, either for agriculture or for timber, tends to leave a highly erodible land surface. In many tropical countries erosion from deforested areas is having a devastating influence on coastal zones including destruction of coral reefs far offshore. Poor land management practices such as overgrazing, especially on hill lands, always leads to serious erosion problems which are difficult or impossible to remedy due to the scale of damage and cost of reconstructing hill-sides. Erosion control measures for semi-arid areas are extensively described by Hudson.

Chapter 5

Fertilizers as Water Pollutants

EUTROPHICATION OF SURFACE WATERS

"Eutrophication" is the enrichment of surface waters with plant nutrients. While eutrophication occurs naturally, it is normally associated with anthropogenic sources of nutrients. The "trophic status" of lakes is the central concept in lake management. It describes the relationship between nutrient status of a lake and the growth of organic matter in the lake. Eutrophication is the process of change from one trophic state to a higher trophic state by the addition of nutrient.

Agriculture is a major factor in eutrophication of surface waters. The most complete global study of eutrophication was the Organization for Economic Cooperation and Development (OECD) Cooperative Programme on Eutrophication carried out in the 1970s in eighteen countries. The sequence of trophic state, from oligotrophic (nutrient poor) to hypertrophic (= hypereutrophic [nutrient rich]).

Although both nitrogen and phosphorus contribute to eutrophication, classification of trophic status usually focuses on that nutrient which is limiting. In the majority of cases, phosphorus is the limiting nutrient.

While the effects of eutrophication such as algal blooms are readily visible, the process of eutrophication is complex and its measurement difficult. This is not the place for a major discussion on the science of eutrophication.

	Phosphorus	*Chlorophyll*
Ultra-oligotrophic	10%	6%
Oligotrophic	63%	49%
Mesotrophic	26%	42%
Eutrophic	1%	3%
Hypertrophic	0%	0%

Because of the complex interaction amongst the many variables that play a part in eutrophication, Janus and Vollenweider (1981) concluded that it is impossible to develop strict boundaries between trophic classes.

They calculated, for example, the probability (as %) of classifying a lake with total phosphorus and chlorophyll–a concentrations of 10 and 2.5 mg/m^3 respectively.

The symptoms and impacts of eutrophication are:

- Increase in production and biomass of phytoplankton, attached algae and macrophytes.
- Shift in habitat characteristics due to change in assemblage of aquatic plants.
- Replacement of desirable fish (e.g., salmonids in western countries) by less desirable species.
- Production of toxins by certain algae.
- Increasing operating expenses of public water supplies, including taste and odour problems, especially during periods of algal blooms.
- Deoxygenation of water, especially after collapse of algal blooms, usually resulting in fish kills.
- Infilling and clogging of irrigation canals with aquatic weeds (water hyacinth is a problem of introduction, not necessarily of eutrophication).
- Loss of recreational use of water due to slime, weed infestation and noxious odour from decaying algae.
- Impediments to navigation due to dense weed growth.
- Economic loss due to change in fish species, fish kills, etc.

ROLE OF AGRICULTURE IN EUTROPHICATION

In their summary of water quality impacts of fertilizers, FAO/ECE cited the following problems:

- Fertilization of surface waters (eutrophication) results in, for example, explosive growth of algae which causes disruptive changes to the biological equilibrium. This is true both for inland waters (ditches, river, lakes) and coastal waters.
- Groundwater is being polluted mainly by nitrates. In all countries groundwater is an important source of drinking water. In several areas the groundwater is polluted to an extent that it is no longer fit to be used as drinking water according to present standards.

While these problems were primarily attributed to mineral fertilizers by FAO/ECE, in some areas the problem is particularly associated with extensive and intensive application of organic fertilizers.

The precise role of agriculture in eutrophication of surface water and contamination of groundwater is difficult to quantify. Where it is warranted, the use of environmental isotopes can aid in the diagnosis of pollutant pathways to and within groundwater. RIVM (1992), citing Isermann (1990), calculated that European agriculture is responsible for 60% of the total riverine flux of nitrogen to the North Sea and 25% of the total phosphorus loading. Agriculture also makes a substantial contribution to the total atmospheric nitrogen loading to the North and the Baltic Seas. This amounts to 65% and 55% respectively. Czechoslovakia reported that agriculture contributes 48% of the pollution of surface water; Norway and Finland reported locally significant eutrophication of surface waters arising from agriculture; high levels of usage of N and P are considered to be responsible for proliferation of algae in the Adriatic; similar observations are made in Danish coastal waters; substantial contamination of groundwater by nitrate in the Netherlands was also reported. Appelgren reported that 50% of shallow groundwater wells that supply more than one million rural residents in Lithuania are unfit for human consumption because of a wide range of pollutants which include pesticides and nitrogen species. In the 1960s Lake Erie was declared "dead" by the press due to the high levels of nutrients accompanied by excessive growth of algae, fish kills and anaerobic bottom sediments.

Although the ECE (1992) regarded livestock wastes as a point source and excluded it from calculations of the contribution of agriculture to eutrophication in Europe, their statistics indicated that livestock wastes accounted "on average" for 30% of the total phosphorus load to European inland waters, with the rest of agriculture accounting for a further 17%. The situation for nitrogen, as for phosphorus, was quite variable from country to country. Danish statistics indicated that manure contributes at least 50% of the leaching of inorganic N. Nitrogen from agricultural non-point sources in the Netherlands amounted to 71% of the total N load generated from within the Netherlands.

A study by Ryding (1986) in Sweden demonstrated how lakes which were unaffected by industrial or municipal point sources, underwent long-term change in nutrient status as a result of agricultural activities in the watershed. Over the period 1973-1981 the nutrient status of Lake Oren increased from 780 to 1000 mg/m^3 for Total-N and from 10 to 45 mg/m^3 for Total-P. Lake transparency declined from 6.2 to 2.6 m and suffered periodic (heavy) algal blooms.

Heavily fertilized crops such as maize tend to have large losses relative to non-intensive uses such as pasture. Agricultural uses associated with poor land management practices that lead to erosion also produce significant nutrient losses. Wastes, manures and sludges, through biological concentration processes, can supply soils with 100 times more hazardous products than fertilizers for the equivalent plant nutrient content. This is considered a major environmental (and water quality) problem in periurban areas of many developing countries. Numerous authors report that a high degree of variability at individual sites is expected as a consequence of changes in hydrological regime from year to year. The implication is that estimation techniques using "typical" values of nutrient yield can expect to have a high degree of uncertainty and could be very much in error if estimated from data collected over a single year.

There is a danger, however, in assuming that all waters have natural levels that are low in nutrients. In some areas, such as lakes located in areas of rich agricultural soils, waters have historically been highly enriched by nutrients associated with natural erosion of fertile soils. In the prairie lakes of Canada, for example, early settlers reported that the lakes were green with algae.

In other parts of the world, as in Asia, ancient civilizations so profoundly impacted water quality that there are no longer "natural" levels of nutrients. In such situations the existence of eutrophication, while undeniable, must be measured against arbitrary standards that reflect water quality criteria established on the basis of societal needs for beneficial use of the water.

ORGANIC FERTILIZERS

The importance and, in some cases, the major problems associated with organic fertilizers, deserve special mention. Manure produced by cattle, pigs and poultry are used as organic fertilizer the world over. To this is added human excreta, especially in some Asian countries where animal and human excreta are traditionally used in fish culture as well as on soils. However, intensive livestock production has produced major problems of environmental degradation, a phenomenon which has been the subject of European and North American legislation and control. The problem is particularly acute in areas of intensive livestock production, such as in the Eastern and Southern parts of the Netherlands where the production of manure greatly exceeds the capacity of the land to assimilate these wastes.

In addition to problems associated with excessive application of manure on the land, is the problem of direct runoff from intensive cattle, pig and poultry farms. Although this is controlled in many western countries, it constitutes a serious problem for water quality in much of the rest of the world. For example, Appelgren reports that discharge of pig wastes from intensive pig raising in Lithuania is a major source of surface water pollution in that country. The FAO/ECE reports similar problems in the Po River of Italy. The Canadian Department of Agriculture calculated in 1978, on the basis of detailed study of several feedlot operations, that cattle feedlots and manure storage facilities contributed 0.5-13% of the total loading of total phosphorus at that time to the Canadian portion of the Lower (agricultural portion) Great Lakes.

To the typical pathways of degradation, that of surface runoff and infiltration into the groundwater, is added the volatilization of ammonia which adds to acidification of land and water. In a review of environmental impacts caused by animal husbandry in Europe, the FAO/ECE (1991) reported the following major categories of impacts:

- Fertilization of surface waters, both as a result of direct discharges of manure and as a consequence of nitrate, phosphate and potassium being leached from the soil.
- Contamination of the groundwater as a result of leaching, especially by nitrate. Phosphates are less readily leached out, but in areas where the soil is saturated with phosphate this substance is found in the groundwater more and more often.
- Surface waters and the groundwater are being contaminated by heavy metals. High concentrations of these substances pose a threat to the health of man and animals. To a certain extent these heavy metals accumulate in the soil, from which they are taken up by crops. For example, pig manure contains significant quantities of copper.
- Acidification as a result of ammonia emission (volatilization) from livestock accommodation, manure storage facilities and manure being spread on the land. Ammonia constitutes a major contribution to the acidification of the environment, especially in areas with considerable intensive livestock arming.

NITROGEN FERTILIZERS **PHOSPHORUS FERTILIZERS**

⇓⇓ ⇓⇓

applied to land surface → Surface Runoff ⇐ applied to land surface

leaching ← leaching

⇒ *Interflow / tile drains*

⇓⇓ ↓ ↓

loss to ground water loss to ground water

Fig:1. Schematic Diagram of Nitrogen and Phosphorus Losses. Arrows are proportional to loss

ENVIRONMENTAL CHEMISTRY

The key hydrological processes that link rainfall, runoff and leaching and which give rise to erosion and transport of chemically enriched soil particles, are important components of the environmental chemistry, transport and fate of fertilizer products. The environmental dynamics of nitrogen and phosphorus are well known although the detailed transformations of nitrogen that occur in soil and water are difficult to study and document.

Nitrogen is comprised of the forms: soluble organic N, NH_4-N (ammonium), NO_3-N (nitrate), NO_2-N (nitrite) and N associated with sediment as exchangeable NH_4-N or organic-N. Nitrogen cycling is extremely dynamic and complex, especially the microbiological processes responsible for mineralization, fixation and denitrification of soil nitrogen. Generally, in soils that are not waterlogged, soil N (held as protein in plant matter) and fertilizer-N are microbiologically transformed to NH_4 (ammonium) through the process of ammonification.

The ammonium ion is oxidized by two groups of bacteria (Nitrosomonas and Nitrobacter) to NO_3 with an unstable intermediate NO_2 product in a process called nitrification. Urea is readily hydrolysed to ammonium. Denitrification occurs under anoxic conditions such as wetlands where NO_3 is reduced to various gaseous forms. The N cycle is largely controlled by bacteria, hence the rate of N cycling is dependent upon factors such as soil moisture, temperature, pH, etc. NO_3 is the end-product of aerobic N decomposition and is always dissolved and mobile.

From a water quality perspective, the ammonium ion (NH_4) can be adsorbed to clay particles and moved with soil during erosion. More importantly, however, NH_4 and NO_3 are soluble and are mobilized through the soil profile to groundwater during periods of rain by the process of leaching. NO_3 is also observed in surface runoff during rainfall events. Prevention of nitrogen pollution of surface and groundwater depends very much on the ability to maintain NO_3 in soil only up to the level that can be taken up by the crop and to reduce the amount of NO_3 held in the soil after harvesting.

In contrast, the behaviour of phosphorus is quite simple. Phosphorus can exist in a variety of forms: as mineral (generally apatite) phosphorus (AP); non-apatite inorganic-P (NAIP); organic-P (OP - bound up with carbon and oxygen in plant matter); and as dissolved soluble reactive ortho-P (SRP). The phosphorus species AP, NAIP and OP are associated with the particulate phase. In studies of phosphorus movement from agricultural lands the largest amount is sorbed onto clay materials and transported as erosion products.

SRP is readily available to aquatic plants to the point where measured SRP in surface water may only represent a residual amount after most of the SRP has been taken up by plant life. Consequently, in aquatic studies, the focus is often on the sediment-associated forms of P as these tend to dominate total phosphorus flux. The NAIP fraction is considered to be available to plant roots and is rapidly solubilized under conditions of anoxia in the bottom of lakes and reservoirs. It is for this reason that lake sediments can represent a very large internal (autotrophic) load of phosphorus which is recycled into the water column during periods of bottom anoxia. This load can be so large that, without attention to lake sediment remediation, phosphorus management programmes in tributaries can be quite meaningless.

The Point Versus Non-point Source Dilemma

The dilemma in many countries is to ascertain the role of agriculture relative to the impacts of (often untreated) municipal sewage. In a large number of countries the database required to make this distinction is lacking and frustrates the development of a rational pollution abatement programme and inhibits cost-effective investment in control measures. In developing countries it makes sense that the focus should initially be on point source control; however, it has been the experience in the developed countries that point source control for nutrients has not had the desired level of environmental benefit until agricultural control measures were seriously addressed.

It is significant that the trend of fertilizer usage worldwide has been one of huge increases in the past 40 years suggesting that, in the absence of major changes in land use to control fertilizer runoff in large parts of the world, one may expect that agriculture will be responsible for an ever-increasing contribution to surface water pollution.

The observations reported by Quirós for the La Plata basin are indicative of the difficulty in segregating the effects of agriculture from other sources. In the Great Lakes of North America some $10 million dollars were spent between 1970 and 1980 to quantify the relative impacts of point versus non-point sources. That exercise proved enormously successful and specific policies were adopted for nutrient control in each lake basin that reflected the relative contributions from each type of source.

MANAGEMENT OF WATER QUALITY IMPACTS FROM FERTILIZERS

The investigation of eutrophication of surface water by agriculture must adopt a pragmatic management perspective. Of value to agriculture is the perspective adopted by the OECD study of eutrophication. That study focused on the following aspects:

- The qualitative assessment of the trophic state of bodies of water in terms of a few easily measured parameters.
- The dependence of this state on nutritional conditions and nutrient load.
- Translation of these results to the needs of eutrophication control for management.

The progression of these aspects is interesting in that the focus is on easily measurable state of the water body, followed by a determination of the extent to which the state is a product of nutrient loads, then the degree to which loads may be manipulated to achieve a desired trophic state that is determined by water use. Prediction of water quality impacts of fertilizers and related land management practices is an essential element of site-specific control options and for the development of generic approaches for fertilizer control. Prediction tools are essentially in the form of models.

Mineral Fertilizers

The response to the need to control leaching and runoff of nutrients and contamination of soils and water by heavy metals has been variable in Europe. Control measures are part of the larger issue of mineral and organic fertilizer usage. FAO/ECE (1991) summarized the types of voluntary and mandated controls in Europe that apply to mineral fertilizers as:

- Taxes on fertilizer.
- Requirement for fertilizer plans.
- Preventing the leaching of nutrients after the growing season by increasing the area under autumn/winter green cover and by sowing crops with elevated nitrogen.
- Promoting and subsidizing better application methods, developing new, environmentally sound fertilizers and promoting soil testing.
- Severely limiting the use of fertilizers in, e.g., water extraction areas and nature protection areas.

In any location where intensive agriculture and/or livestock farming produces serious risks of nitrogen pollution, Ignazi (1993) recommended the following essential steps that are taken at the farm level:

Rational Nitrogen Application

To avoid over-fertilization, the rate of nitrogen fertilizer to be applied needs to be calculated on the basis of the "crop nitrogen balance". This takes into account plant needs and amount of N in the soil.

Vegetation Cover

As far as possible, keep the soil covered with vegetation. This inhibits build-up of soluble nitrogen by absorbing mineralized nitrogen and preventing leaching during periods of rain.

Manage the Period between Crops

Organic debris produced by harvesting is easily mineralized into leachable N. Steps to reduce leachable N includes planting of "green manure" crops and delaying ploughing of straw, roots and leaves into the soil.

Rational Irrigation

Poor irrigation has one of the worst impacts on water quality, whereas precision irrigation is one of the least polluting practices as well as reducing net cost of supplied water.

Optimize other Cultivation Techniques

Highest yields with minimum water quality impacts require optimization of practices such as weed, pest and disease control, liming, balanced mineral fertilizers including trace elements, etc.

Agricultural Planning

Implement erosion control techniques that complement topographic and soil conditions.

Organic Fertilizers

Voluntary and legislated control measures in Europe are intended to have the following benefits:

- Reduce the leaching of nutrients.
- Reducing emissions of ammonia.
- Reducing contamination by heavy metals.

The nature of these measures varies from country to country; however, FAO/ECE have summarized the types of voluntary and mandated control as:

- Maximum numbers of animals per hectare based on amount of manure that can be safely applied per hectare of land.
- Maximum quantities of manure that can be applied on the land is fixed, based on the N and P content of the manure.
- Holdings wishing to keep more than a given number of animals must obtain a licence.
- The periods during which it is allowed to apply manure to the land have been limited and it is obligatory to work it into the ground immediately afterwards.
- Establishment of regulations on minimum capacity for manure storage facilities.
- Establish fertilizer plans.
- Levies (taxes) on surplus manure.
- Areas under autumn/winter green cover were extended and green fallowing is being promoted.

- Maximum amounts established for spreading of sewage sludge on land based on heavy metal content.
- Change in composition of feed to reduce amount of nutrients and heavy metals.
- Research and implementation of means of reducing ammonia loss.

Sludge Management

Sludge is mentioned here only insofar as the spreading of sludge from municipal wastewater treatment facilities on agricultural land is one method used to get rid of municipal sludge in a way that is perceived to be beneficial. The alternatives are incineration and land fill. FAO/ECE include sludge within the category of organic fertilizers but note that sludge often contains unacceptable levels of heavy metals. Pollution of water by sludge runoff is otherwise the same as for manure noted above.

ECONOMICS OF CONTROL OF FERTILIZER RUNOFF

Nutrient loss is closely associated with rainfall-runoff events. For phosphorus, which tends to be associated with the solid phase (sediment), runoff losses are directly linked to erosion. Therefore the economics of nutrient control tend to be closely tied to the costs of controlling runoff and erosion. Therefore, this will be treated briefly here. In particular, it is useful to examine the economic cost of nutrient runoff which must be replaced by fertilizers if the land is to remain productive.

The link between erosion, increasing fertilizer application and loss of soil productivity is very direct in many countries. In the Brazilian state of Paraná where agriculture is the base of the state economy, Paraná produces 22% of the national grain production on only 2.4% of the Brazilian territory. Agricultural expansion in Paraná occurred mainly in the period 1950-1970 and was "characterized by short-term agricultural systems leading to continuous and progressive environmental degradation as a result of economic policies and a totally inappropriate land parcelling and marketing system...". Erosion has led to extensive loss of top soil, large-scale gullying and silting of ditches and rivers. The use of fertilizers has risen as a consequence, up 575% over the period 1970-1986 and without any gain in crop yields. Loss of N-P-K from an average erosion of 20 t/ha/yr represents an annual economic loss of US$242 million in nutrients.

Analysis by Elwell and Stocking (1982) of nutrient loss arising from erosion in Zimbabwe shows similar significant economic losses in African situations. Stocking, applying data collected in the 1960s by Hudson to the soil use map of Zimbabwe, calculated an annual loss of 10 million tonnes of nitrogen and 5 million tonnes of phosphorus annually as a consequence of erosion. Roose also cites losses of 98 kg/ha/yr of nitrogen, 29 kg/ha/yr of phosphorus, 39 kg/ha/yr of lime and 39 kg/ha/yr of magnesium from soils of lower Côte d'Ivoire as a result of erosion. This loss is so severe that compensation requires 7 tonnes of fresh manure annually, plus 470 kg of ammonium sulphate, 160 kg of superphosphate, 200 kg dolomite and 60 kg of potassium chloride per hectare per year. Roose notes that it is not surprising that the soil is exhausted after only two years of traditional agriculture. Furthermore, these losses do not take into account additional losses due to harvesting and runoff. Roose summarizes by stating that action against soil erosion is essential in order to manage what he describes as a "terrible" chemical imbalance in soils caused by soil erosion.

Economic losses tend to be higher in tropical countries where soils, rainfall and agricultural practices are more conducive to erosion and reported rates of erosion are much above average. The World Bank (1992) reported that extrapolation from test-plots of impacts of soil loss on agricultural productivity, indicates some 0.5-1.5% loss of GDP annually for countries such as Costa Rica, Malawi, Mali and Mexico. These losses do not include offsite costs such as reservoir infilling, river sedimentation, damage to irrigation systems, etc.

Soil fertility is a complex issue and nutrient loss is not necessarily nor always a consequence of erosion. Erosion and soil loss is the end member of a variety of physical, vegetative and nutrient factors that lead to soil degradation. Global patterns of fertilizer application, as reported by Joly (1993), indicate however that rapidly rising levels of fertilizer utilization are required merely to maintain soil productivity from a variety of types of loss, including losses due to erosion and, more generally, to soil degradation.

In a study of 17 agricultural sub-watersheds in the Lake Balaton district of Hungary, Jolankai (1986) measured and modelled N and P runoff from a variety of agricultural land uses. He calculated that a selection of control measures (mainly erosion control) would reduce phosphorus loss by 52.8% at a cost of US$ 2500 per ha in remediation measures (in 1986).

AQUACULTURE

Aquaculture is a special case of agricultural pollution. There are two main forms: land-based and water-based systems. Effluent controls are possible on land-based systems, however, water-based systems present particular problems. Aquaculture is rapidly expanding in most parts of the developed and developing world, both in freshwater and marine environments. In contrast, coastal fisheries in most countries are declining.

The environmental impact is primarily a function of feed composition and feed conversion (faecal wastes), plus assorted chemicals used as biocides, disinfectants, medicines, etc. Wastage of feed (feed not taken up by the fish) is estimated to be 20% in European aquaculture. Waste feed and faecal production both add substantial nutrient loadings to aquatic systems.

Additional environmental problems include risk of disease and disease transfer to wild fish, introduction of exotic species, impacts on benthic communities and on the eutrophication of water, interbreeding of escaped cultured fish with wild fish with consequent genetic change in the wild population.

Traditional integrated aquaculture systems, as in China, where sewage-fish culture is practised, can be a stabilizing influence in the entire ecosystem. This is recommended, especially in developing countries where water and resources are scarce or expensive.

PROBLEMS OF RESTORATION OF EUTROPHIC LAKES

Eutrophic and hypertrophic lakes tend to be shallow and suffer from high rates of nutrient loadings from point and non-point sources. In areas of rich soils such as the Canadian prairies, lake bottom sediments are comprised of nutrient-enriched soil particles eroded from surrounding soils. The association of phosphorus with sediment is a serious problem in the restoration of shallow, enriched lakes. P-enriched particles settle to the bottom of the lake and form a large pool of nutrient in the bottom sediments that is readily available to rooted plants and which is released from bottom sediments under conditions of anoxia into the overlying water column and which is quickly utilized by algae.

This phosphorus pool, known as the "internal load" of phosphorus, can greatly offset any measures taken by river basin managers to control lake eutrophication by control of external phosphorus sources from agriculture and from point sources. Historically, dredging of bottom sediments was considered the only means of remediating nutrient-rich lake sediments, however, modern technology now provides alternative and more cost-effective methods of controlling internal loads of phosphorus by oxygenation and by chemically treating sediments *in situ* to immobilize the phosphorus. Nevertheless, lake restoration is expensive and must be part of a comprehensive river basin management programme.

PESTICIDES AS WATER POLLUTANTS

The term "pesticide" is a composite term that includes all chemicals that are used to kill or control pests. In agriculture, this includes herbicides (weeds), insecticides (insects), fungicides (fungi), nematocides (nematodes) and rodenticides (vertebrate poisons). A fundamental contributor to the Green Revolution has been the development and application of pesticides for the control of a wide variety of insectivorous and herbaceous pests that would otherwise diminish the quantity and quality of food produce.

The use of pesticides coincides with the "chemical age" which has transformed society since the 1950s. In areas where intensive monoculture is practised, pesticides were used as a standard method for pest control. Unfortunately, with the benefits of chemistry have also come disbenefits, some so serious that they now threaten the long-term survival of major ecosystems by disruption of predator-prey relationships and loss of biodiversity. Also, pesticides can have significant human health consequences. While agricultural use of chemicals is restricted to a limited number of compounds, agriculture is one of the few activities where chemicals are intentionally released into the environment because they kill things. Agricultural use of pesticides is a subset of the larger spectrum of industrial chemicals used in modern society.

The American Chemical Society database indicates that there were some 13 million chemicals identified in 1993 with some 500 000 new compounds being added annually. In the Great Lakes of North America, for example, the International Joint Commission has estimated that there are more than 200 chemicals of concern in water and sediments of the Great Lakes ecosystem.

Because the environmental burden of toxic chemicals includes both agriculture and non-agricultural compounds, it is difficult to separate the ecological and human health effects of pesticides from those of industrial compounds that are intentionally or accidentally released into the environment. However, there is overwhelming evidence that agricultural use of pesticides has a major impact on water quality and leads to serious environmental consequences.

Irrigated agriculture, especially in tropical and subtropical environments, usually requires modification of the hydrological regime which, in turn, creates habitat that is conducive to breeding of insects such as mosquitoes which are responsible for a variety of vector-borne diseases. In addition to pesticides used in the normal course of irrigated agriculture, control of vector-borne diseases may require additional application of insecticides such as DDT which have serious and widespread ecological consequences. In order to address this problem, environmental management methods to control breeding of disease vectors are being developed and tested in many irrigation projects.

Historical Development of Pesticides

The history of pesticide development and use is the key to

understanding how and why pesticides have been an environmental threat to aquatic systems and why this threat is diminishing in developed countries and remains a problem in many developing countries.

North-South Dilemma Over Pesticide Economics

As noted above, the general progression of pesticide development has moved from highly toxic, persistent and bioaccumulating pesticides such as DDT, to pesticides that degrade rapidly in the environment and are less toxic to non-target organisms. The developed countries have banned many of the older pesticides due to potential toxic effects to man and/or their impacts on ecosystems, in favour of more modern pesticide formulations. In the developing countries, some of the older pesticides remain the cheapest to produce and, for some purposes, remain highly effective as, for example, the use of DDT for malaria control. Developing countries maintain that they cannot afford, for reasons of cost and/or efficacy, to ban certain older pesticides. The dilemma of cost/efficacy versus ecological impacts, including long range impacts via atmospheric transport and access to modern pesticide formulations at low cost remains a contentious global issue.

In addition to ecological impacts in countries of application, pesticides that have been long banned in developed countries (such as DDT, toxaphene, etc.), are consistently found in remote areas such as the high arctic. Chemicals that are applied in tropical and subtropical countries are transported over long distances by global circulation. The global situation has deteriorated to the point where many countries are calling for a global convention on "POPs" (Persistent Organic Pollutants) which are mainly chlorinated compounds that exhibit high levels of toxicity, are persistent and bioaccumulate. The list is not yet fixed; however, "candidate" substances include several pesticides that are used extensively in developing countries.

FATE AND EFFECTS OF PESTICIDES

Human Health Effects of Pesticides

Perhaps the largest regional example of pesticide contamination and human health is that of the Aral Sea region. UNEP (1993) linked the effects of pesticides to "the level of oncological (cancer), pulmonary and haematological morbidity, as well as on inborn deformities... and immune system deficiencies".

Human health effects are caused by:

- *Skin contact:* Handling of pesticide products.
- *Inhalation:* Breathing of dust or spray.

- *Ingestion:* Pesticides consumed as a contaminant on/in food or in water.

Farm workers have special risks associated with inhalation and skin contact during preparation and application of pesticides to crops. However, for the majority of the population, a principal vector is through ingestion of food that is contaminated by pesticides. Degradation of water quality by pesticide runoff has two principal human health impacts. The first is the consumption of fish and shellfish that are contaminated by pesticides; this can be a particular problem for subsistence fish economies that lie downstream of major agricultural areas.

The second is the direct consumption of pesticide-contaminated water. WHO (1993) has established drinking water guidelines for 33 pesticides. Many health and environmental protection agencies have established "acceptable daily intake" (ADI) values which indicate the maximum allowable daily ingestion over a person's lifetime without appreciable risk to the individual. For example, in a recent paper by Wang and Lin (1995) studying substituted phenols, tetrachlorohydroquinone, a toxic metabolite of the biocide pentachlorophenol, was found to produce "significant and dose-dependent DNA damage".

Ecological Effects of Pesticides

Pesticides are included in a broad range of organic micro pollutants that have ecological impacts. Different categories of pesticides have different types of effects on living organisms, therefore generalization is difficult. Although terrestrial impacts by pesticides do occur, the principal pathway that causes ecological impacts is that of water contaminated by pesticide runoff. The two principal mechanisms are bioconcentration and biomagnification.

Bioconcentration

This is the movement of a chemical from the surrounding medium into an organism. The primary "sink" for some pesticides is fatty tissue ("lipids"). Some pesticides, such as DDT, are "lipophilic", meaning that they are soluble in and accumulate in, fatty tissue such as edible fish tissue and human fatty tissue. Other pesticides such as glyphosate are metabolized and excreted.

Biomagnification

This term describes the increasing concentration of a chemical as food energy is transformed within the food chain. As smaller organisms are eaten by larger organisms, the concentration of pesticides and other chemicals are increasingly magnified in tissue and other organs. Very high concentrations can be observed in top predators, including man. The ecological effects of pesticides (and other organic contaminants) are varied and are often inter-related. Effects at the organism or ecological level are usually considered to be an early warning indicator of potential human health impacts.

The major types of effects are listed below and will vary depending on the organism under investigation and the type of pesticide. Different pesticides have markedly different effects on aquatic life which makes generalization very difficult. The important point is that many of these effects are chronic (not lethal), are often not noticed by casual observers, yet have consequences for the entire food chain.

- Death of the organism.
- Cancers, tumours and lesions on fish and animals.
- Reproductive inhibition or failure.
- Suppression of immune system.
- Disruption of endocrine (hormonal) system.
- Cellular and DNA damage.
- Teratogenic effects (physical deformities such as hooked beaks on birds).
- Poor fish health marked by low red to white blood cell ratio, excessive slime on fish scales and gills, etc.
- Intergenerational effects (effects are not apparent until subsequent generations of the organism).
- Other physiological effects such as egg shell thinning.

These effects are not necessarily caused solely by exposure to pesticides or other organic contaminants, but may be associated with a combination of environmental stresses such as eutrophication and pathogens.

These associated stresses need not be large to have a synergistic effect with organic micro pollutants. Ecological effects of pesticides extend beyond individual organisms and can extend to ecosystems. Swedish work indicates that application of pesticides is thought to be one of the most significant factors affecting biodiversity. Jonsson report that the continued decline of the Swedish partridge population is linked to changes in land use and the use of chemical weed control.

Chemical weed control has the effect of reducing habitat, decreasing the number of weed species and of shifting the balance of species in the plant community. Swedish studies also show the impact of pesticides on soil fertility, including inhibition of nitrification with concomitant reduced uptake of nitrogen by plants. These studies also suggest that pesticides adversely affect soil micro-organisms which are responsible for microbial degradation of plant matter (and of some pesticides) and for soil structure.

NATURAL FACTORS THAT DEGRADE PESTICIDES

In addition to chemical and photochemical reactions, there are two principal biological mechanisms that cause degradation of pesticides. These are: (1) microbiological processes in soils and water and (2) metabolism of pesticides that are ingested by organisms as part of their food supply. While both processes are beneficial in the sense that pesticide toxicity is reduced, metabolic processes do cause adverse effects in, for example, fish. Energy used to metabolize pesticides and other xenobiotics (foreign chemicals) is not available for other body functions and can seriously impair growth and reproduction of the organism.

Degradation of Pesticides in Soil

Many pesticides dissipate rapidly in soils. This process is mineralization and results in the conversion of the pesticide into simpler compounds such H_2O, CO_2 and NH_3. While some of this process is a result of chemical reactions such as hydrolysis and photolysis, microbiological catabolism and metabolism is usually the major route of mineralization. Soil microbiota utilize the pesticide as a source of carbon or other nutrients. Some chemicals (for example 2,4-D) are quite rapidly broken down in soil while others are less easily attacked (2,4,5-T). Some chemicals are very persistent and are only slowly broken down (atrazine).

Process of Metabolism

Metabolism of pesticides in animals is an important mechanism by which organisms protect themselves from the toxic effects of xenobiotics (foreign chemicals) in their food supply. In the organism, the chemical is transformed into a less toxic form and either excreted or stored in the organism. Different organs, especially the liver, may be involved, depending on the chemical. Enzymes play an important role in the metabolic process and the presence of certain enzymes, especially "mixed" function oxygenases (MFOs) in liver, is now used as an indicator that the organism has been exposed to foreign chemicals.

PESTICIDE MONITORING IN SURFACE WATER

Monitoring data for pesticides are generally poor in much of the world and especially in developing countries. Key pesticides are included in the monitoring schedule of most Western countries, however, the cost of analysis and the necessity to sample at critical times of the year (linked to periods of pesticide use) often preclude development of an extensive data set. Many developing countries have difficulty carrying out organic chemical analysis due to problems of inadequate facilities, impure reagents and financial constraints. New techniques using immunoassay procedures for presence/absence of specific pesticides may reduce costs and increase reliability. Immunoassay tests are available for triazines, acid amides, carbamates, 2,4-D/phenoxy acid, paraquot and aldrin

Data on pesticide residues in fish for lipophilic compounds and determination of exposure and/or impact of fish to lipophobic pesticides through liver and/or bile analysis is mainly restricted to research programmes. Hence, it is often difficult to determine the presence, pathways and fate of the range of pesticides that are now used in large parts of the world. In contrast, the ecosystemic impacts from older, organochlorine pesticides such as DDT, became readily apparent and has resulted in the banning of these compounds in many parts of the world for agricultural purposes.

Older pesticides, together with other hydrophobic carcinogens such as PAHs and PCBs, are poorly monitored when using water samples. As an example, the range of concentration of suspended solids in rivers is often between 100 and 1000 mg/l except during major runoff events when concentrations can greatly exceed these values. Tropical rivers that are unimpacted by development have very low suspended sediment concentrations, but increasingly these are a rarity due to agricultural expansion and deforestation in tropical countries. As an example, approximately 67% of DDT is transported in association with suspended matter at sediment concentrations as low as 100 mg/l and increases to 93% at 1000 mg/l of suspended sediment.

Given the analytical problems of inadequate detection levels and poor quality control in many laboratories of the developing countries, plus the fact that recovery rates vary from 50-150% for organic compounds, it follows that monitoring data from water samples are usually a poor indication of the level of pesticide pollution for compounds that are primarily associated with the solid phase. The number of NDs (Not Detectable) in many databases is almost certainly an artifact of the wrong sampling medium (water) and, in some cases, inadequate analytical facilities and procedures.

Clearly, this makes pesticide assessment in water difficult in large parts of the world. Experience suggests that sediment-associated pesticide levels are often much higher than recorded and NDs are often quite misleading. Some water quality agencies now use multi-media (water + sediment + biota) sampling in order to more accurately characterize pesticides in the aquatic environment.

Pesticide monitoring requires highly flexible field and laboratory programmes that can respond to periods of pesticide application, which can sample the most appropriate medium (water, sediment, biota), are able to apply detection levels that have meaning for human health and ecosystem protection and which can discriminate between those pesticides which appear as artifacts of historical use versus those that are in current use. For pesticides that are highly soluble in water, monitoring must be closely linked to periods of pesticide use. In the United States where there have been major studies of the behaviour of pesticide runoff, the triazines (atrazine and cyanazine) and alachlor (chlorinated acetamide) are amongst the most widely used herbicides.

These are used mainly in the spring Studies by Schottler indicate that 55-80% of the pesticide runoff occurred in the month of June. The significance for monitoring is that many newer and soluble pesticides can only be detected shortly after application; therefore, monitoring programmes that are operated on a monthly or quarterly basis (typical of many countries) are unlikely to be able to quantify the presence or determine the significance of pesticides in surface waters. Pesticides that have limited application are even less likely to be detected in surface waters. The danger lies in the presumption by authorities that ND (non-detectable) values implies that pesticides are absent. It may well only mean that monitoring programmes failed to collect data at the appropriate times or analysed the wrong media.

PESTICIDE MANAGEMENT AND CONTROL

Prediction of water quality impacts of pesticides and related land management practices is an essential element of site-specific control options and for the development of generic approaches for pesticide control. Prediction tools are mainly in the form of models. Also, the key hydrological processes that control infiltration and runoff and erosion and sediment transport, are controlling factors in the movement of pesticides.

The European Experience

The Netherlands National Institute of Public Health and Environmental Protection concluded that "groundwater is threatened by pesticides in all European states. This is obvious both from the available monitoring data and calculations concerning pesticide load, soil sensitivity and leaching.... It has been calculated that on 65% of all agricultural land the EC standard for the sum of pesticides (0.5; m g/l) will be exceeded. In approximately 25% of the area this standard will be exceeded by more than 10 times...."

In recognition of pesticide abuse and of environmental and public health impacts the European countries have adopted a variety of measures that include the following:

- Reduction in use of pesticides (by up to 50% in some countries).
- Bans on certain active ingredients.

- Revised pesticide registration criteria.
- Training and licensing of individuals that apply pesticides.
- Reduction of dose and improved scheduling of pesticide application to more effectively meet crop needs and to reduce preventative spraying.
- Testing and approval of spraying apparatus.
- Limitations on aerial spraying.
- Environmental tax on pesticides.
- Promote the use of mechanical and biological alternatives to pesticides.

Elsewhere, as for example Indonesia, reduction in subsidies has reduced the usage of pesticides and has increased the success of integrated pesticide management programmes.

Pesticide Registration

Pesticide control is mainly carried out by a system of national registration which limits the manufacture and/or sale of pesticide products to those that have been approved. In developed countries, registration is a formal process whereby pesticides are examined, in particular, for mammalian toxicity (cancers, teratogenic and mutagenic effects, etc.) and for a range of potential environmental effects based on the measured or estimated environmental behaviour of the product based on its physico-chemical properties.

Most developing countries have limited capability to carry out their own tests on pesticides and tend to adopt regulatory criteria from the developed world. As our knowledge of the effects of pesticides in the environment accumulates, it has become apparent that many of the older pesticides have inadequate registration criteria and are being re-evaluated. As a consequence, the environmental effects of many of the older pesticides are now recognized as so serious that they are banned from production or sale in many countries.

A dilemma in many developing countries is that many older pesticides (e.g., DDT) are cheap and effective. Moreover, regulations are often not enforced with the result that many pesticides that are, in fact, banned, are openly sold and used in agricultural practice. The dichotomy between actual pesticide use and official policy on pesticide use is, in many countries, far apart.

Regulatory control in many countries is ineffective without a variety of other measures, such as education, incentives, etc. The extent to which these are effective in developed versus developing countries depends very much on:

- The ability of government to effectively regulate and levy taxes.

- On the ability or readiness of the farming community to understand and act upon educational programmes.

The fundamental dilemma remains one of accommodating local and short term gain by the farmer (and manufacturer and/or importer) by application of an environmentally dangerous pesticides, with societal good by the act of limiting or banning its use.

There is now such concern over environmental and, in some instances, human health effects of excessive use and abuse of pesticides, that there is active discussion within many governments of the need to include a programme of pesticide reduction as part of a larger strategy of sustainable agriculture. In 1992, Denmark, the Netherlands and Sweden were the first of the 24 member states of the OECD to embark upon such a programme. The Netherlands is the world's second largest exporter of agricultural produce after the United States.

In contrast, wood preservatives in the forest sector account for 70% of Swedish pesticide use with agriculture using only 30%. As noted above, the lack of baseline data on pesticides in surface waters of OECD countries, is a constraint in establishing baseline values against which performance of the pesticide reduction programme can be measured.

Chapter 6

Irrigation Service Supply

THE INFRASTRUCTURE

Capital and Current Spending

The irrigation cycle's supply side consists of appropriating the five base water flows, their storage and distribution to and from storage through to farmers' fields. The economics of irrigation service supply– first, the making of the infrastructure and, second, the year-on-year delivery of water to farming families and agribusinesses. The Report of the World Commission on Dams (WCD) has made clear just how ancient are the infrastructural works of irrigation:

> The earliest evidence of river engineering is the ruins of irrigation canals over eight thousand years old in Mesopotamia. Remains of water storage dams found in Jordan, Egypt and other parts of the Middle East date back to at least 3000 BC. Historical records suggest that the use of dams for irrigation and water supply became more widespread about a thousand years later. At that time, dams were built in the Mediterranean region, China and Meso America. Remains of earth embankment dams built for diverting water to large community reservoirs can still be found in Sri Lanka and Israel. The Dujiang irrigation project, which supplied 800,000 hectares in China, is 2,200 years old.

The irrigation cycle draws on an immense variety of the economy's real resources both for infrastructural provision and for the subsequent annual operation and maintenance of the service. These resources include:

- Freshwater or brackish water at the point of appropriation.
- The land sites required for the system's infrastructure.
- Wells, boreholes, pumps and their power units such as combustion engines and electric motors.
- Reservoirs.
- Plant for mechanical, biological and chemical treatment of wastewater when treated wastewater is used for irrigation.

- An extensive system of canals, pipes and other channels.
- A wide variety of monitoring, measurement and control devices.
- Petrol and diesel fuel, electricity supplies and spare parts.
- The human resources necessary to design, build, operate, monitor, maintain, repair, rehabilitate, manage and administer the whole.

Headworks costs are those of abstraction and storage. Network costs are those of water distribution. In the analysis of the supply of irrigation services, the economist converts the tabulation of these real resources into money expenditures, on the basis of the resources' market prices. Expenditure can then be divided into capital account spending and current account spending.

Capital expenditure is defined as that where the resource purchased has an expected life of more than 12 months. This would include: land; civil engineering infrastructures such as boreholes, reservoirs, works road networks, pumps and pipes; buildings and plant of all kinds; and vehicles. The real resource requirements for infrastructural supply cross-classified by the three basic supply-side activities of the irrigation cycle.

Current expenditure refers to purchased resources that are either immediately used up in the process of production or which have a life of 12 months or less. These include materials, chemicals, spare parts, electric power and labour time. In practice, relatively small expenditures, such as those on some types of office equipment, would be counted as current expenditure, even where their life exceeds 1 year.

Where long-life resources are rented rather than purchased, or in other instances where payment is made on a regular basis rather than as a lump sum, the expenditure will be classified as a current account item. The rent of land and buildings, for example, will appear under current expenditure. From an economic and a financial perspective, the importance of the distinction between capital and current spending is that, *first,* investment in a newly developed system's civil engineering infrastructures and plant requires far greater expenditure on capital account than the current account costs of a single year's operations. As a result, capital spending often requires debt finance.

Second, commitment to current account spending is more flexible than the sunk costs under capital account. *Third,* the balance between capital and current spending is of the greatest importance in project evaluation. *Fourth,* the profit and loss account of an agribusiness is made up only of current account incomes and expenditures. The profit and loss tabulation is the professional accountant's version of the simpler farm budget.

It has to be admitted that the current-capital distinction may be fuzzy. For example, the patching of a canal with low-quality concrete mortar to plug cracks doubtless would be classified as a current cost. But the replacement of a significant length of canal line with pukka concrete intended to last 10 years would fall under capital spending.

Real capital resource requirements for infrastructural supply Appropriation of the five base flows Land sites; wells and boreholes; *qanat/falaj* system headworks; surface-water off-take structures; flood recession works; positive displacement pumps; roto-dynamic pumps; power units; treatment works for waste water appropriated for supply; measurement and control devices; miscellaneous buildings, roads, vehicles and equipment; weed screens; gantry-mounted grabs; transformer units

Storage Land sites; aquifers for artificial recharge; reservoirs, spillways, stilling basins, siphons; water tanks and cisterns; measurement and control devices; miscellaneous buildings, roads, vehicles and equipment

Distribution Land sites; buried and surface pipes; lined and unlined canals; sluice-gates; measurement and control devices including cross-regulators, weirs, flumes, drop structures, stilling basins, siphons; hydraulic excavators, bulldozers, tractor-mounted mowers, draglines and skips; miscellaneous buildings, roads, vehicles and equipment.

The Short Term Economies of Capacity Utilization

Efficiency in the appropriation, storage, distribution and use of water for agriculture. In all cases but one these are *output efficiency* measures, having a crop output or crop sales measure as the numerator and a measure of irrigation water supply as the denominator.

Of the first seven efficiency measures, only one is a supply-side ratio. This is E_2-net irrigation requirement divided by the base supply. *Supply-side efficiency* in terms of the supply cost per cubic metre of water delivered to the farmer's field. It is also important to remember that irrigation supply in cubic metres can be measured at many points. An irrigation engineer writes.

In big systems there are very substantial losses before the water gets to the irrigation district boundary; there are more between that and the farm boundary and, where farms are big and open channel distribution systems in use, more losses before the water gets to the field. Finally, for the purpose of comparing alternatives for on-farm delivery, one should measure the supply both at the field boundary and at the plant.

The two measures most often used in this text are S_b, base supply at the start of the irrigation cycle and S_f, the volume delivered to the farmer's fields.

The most important concepts in the irrigation economist's tool-kit to explore supply-side efficiency are *cost functions*. These quantitative relationships describe the cost of supplying output in any time-period at each scale of output from zero units up to the system's theoretical capacity. The ability to shift from one delivery level of irrigation water up to a higher level clearly depends on the time available to make the change. It was Alfred Marshall, one of the greatest British economists, who introduced periodization into the analysis of supply and demand. He distinguished between the short term and the long term. In its application to the irrigation cycle, this simple, twofold distinction can be represented in the following way: in the short term, increased daily output is possible through operational changes or by organizational innovations demanding new procedures, both of which are relatively straightforward in their introduction; in the long term, additional infrastructure is required either in new projects or for the expansion of existing capacity, with associated financial, resource and organizational planning.

In conventional analysis it is possible, for either the short or long term, to calculate the *total cost* of supply at each level of output per day. *Average cost* at each level of output can be derived from this by dividing the total cost by the number of units produced. The *marginal cost* of the *n*th unit of output is defined as the difference in total cost between producing *n* and *(n-1)* units. A related but alternative exposition is set out below. It has the advantage that the terms refer to engineering categories, making it easier for non-economists to grasp quickly. It is also more appropriate than the conventional analysis, since it is a better representation of the way irrigation institutions actually think about their costs. Expenditures are defined using three categories-the parallel with farming family and agribusiness costs:

- Prime costs are defined as those used up in the daily supply of irrigation services and consist of the salaries and wages of the workforce, the costs of power, materials, spare parts and other consumables such as bought-in specialist inputs.
- Overhead costs of production are non-prime costs and are often assumed to be invariant with the level of supply. In the simplest case, where capital expenditure on land, infrastructure and plant has been funded 100 per cent by loans, this element of annual overhead costs would be set equal to the annual interest payable and the principal repayable on the debt incurred. Where capital investment is funded, at least in part, from an irrigation authority's surplus of water charges over prime costs, overhead costs include the amortization charges necessary in the long run to contribute to the rehabilitation and replacement of elements of the system infrastructure. The rent of land and buildings and the leasing costs of capital equipment such as vehicles, are also included in overheads.

- Total costs per year are equal to prime plus overhead costs.

For each level of output, one can calculate overhead cost per unit of output and this function, called average overhead cost, is represented for a hypothetical case, *illustrating the short-term situation*. Average overhead costs fall (at a decelerating rate) as the unchanged level of cost is divided by evergreater levels of production.

So in the short term, supply-side efficiency in terms of the overhead costs of irrigation water supply, i.e. the supply costs of the headworks and networks infrastructure per unit of water delivered, is achieved by operating the irrigation system at its full capacity. In other words, at scales of delivery less than full capacity, the average cost of the appropriation, storage and distribution infrastructure necessary to supply irrigation water rises exponentially.

Where an irrigation system is built to meet the peak period of demand-which may be only 10 days long-the system is operating at below capacity for most of the irrigation season. Alternatives are to design a system that does *not* meet peak demand or to use volumetric pricing to suppress the peak demand for water. In this respect Carruthers and Clark showed an interesting link between surface-water and groundwater abstraction:

> Tubewells may be used where surface canal water is at present supplied to augment basic surface supplies and increase cropping intensity. Augmentation to meet relatively short periods of peak demand may be particularly valuable. The most profitable cropping pattern is unlikely to have an even profile of water demand over the year and a canal system is limited in its capacity to cope with peaks.

The Long Term Economies of Scale

We now switch to the long term. Again our interest is in how average overhead costs vary with output scale. In this case output is increased not by raising the volume of irrigation water supplied towards 100 per cent utilization of *a fixed* infrastructural capacity, but by *adding* to the infrastructural capacity through specified engineering works. We can say economies of scale exist where higher levels of output are associated with lower overhead cost per unit of irrigation water supplied.

For the long term, where output rises because of capacity increases resulting from the investment process, the best approach in understanding scale economies is through *ex ante* calculations, i.e., a view of future possibilities not yet realized–*a planning approach*. The starting point is a real catchment in a specific year, with all the infrastructural investment in irrigation services inherited from the past in place. The interest here is in average overhead cost at each alternative value of a set of supply volumes derived from additions to capacity, from a minimal increment through to any upper limit that one believes to be appropriate. Note that the curve does not consist of a set of points representing how average overhead cost changes as the scale of production expands (or contracts) over time. It is not an evolutionary function. What it represents, for each output level, is the average overhead cost if, in a single leap, capacity were expanded up to that level. It is referred to as a long–term supply function not because the figure represents economies of scale with the passage of time but because the long term is the conventional time-scale over which additions to capacity can take place, in contrast to the operational changes of the short term.

This is worth restating. Average overhead cost variations with output in the short run represent the cost-output relationship as management in real-time shifts output upwards or downwards from one level to another. Average overhead cost variations with output in the long run represent the cost-output relationship for each of a series of separate, mutually exclusive choices. Here it is logically impossible for there to be real-time shifts of output upwards or downward from one level to another. So, in the short term, shifts over time in output levels are represented; but in the long term, mutually exclusive choices between future additions to capacity and to output are represented. The hypothetical function clear economies of scale up to 140 units of output per day. However, here, diseconomies of scale appear above 150 units per day. This cost function is quadratic in nature, i.e., well represented by an equation of the form:

$$\text{AOC} = L^2 + bL + c$$

where AOC is average overhead cost and L is the output quantity the equation is:

$$\text{AOC} = 0.004L^2 - 0.998L + 93.5$$

How can one explain long term economies of scale in irrigation service provision? It is the measure of what economists widely refer to as the *indivisibility* of infrastructural provision. This neatly expresses the idea that, to secure even modest levels of output, major works are necessary–a canal is a good example-and these are negligibly less costly than works that secure more substantial output levels. As a result, AOC falls markedly. The existence of significant scale economies at low- to medium-output levels has a most important social effect. It implies that irrigation service supply constitutes what is widely called a natural monopoly.

The same can hold true in infrastructural provision for fresh and wastewater services in general, as well as for transport, energy supply and telecommunications. The proposition that long term economies of scale exist in the supply of irrigation services should be used more as a working hypothesis than as an accepted fact. Moreover, one can expect variation in the existence of such economies between the infrastructural categories of appropriation, storage and distribution.

With respect to appropriation headworks, the existence or not of economies surely varies between the five base flows of the irrigation cycle. *Rainfall collection* using a harvesting surface and a cistern for storage can work well at the small scale and is used in many parts of the world by farming families for small plot cultivation, particularly for their subsistence needs. We saw such an example in the case study of the small agribusiness in Cornwall, where rainfall collection contributed one-sixth of the farm's irrigation requirements.

Irrigation water supplied by *the reuse of domestic and industrial wastewater* certainly can be expected to exhibit long term scale economies in respect of any wastewater treatment prior to reuse. However, if the wastewater treatment plant's (WWTP) cost is accounted for as part of the domestic and industrial supply system, the average overhead cost of reuse water appropriation is zero.

The reuse in farming of *agricultural drainage water* by means of pumps set in drainage channels seems to operate very effectively at small scales, so no long term scale economies can be expected here.

Surface-water and groundwater abstraction takes place on a wide variety of scales as we have already seen in the Tamil Nadu, Curu Valley and Cornwall case studies. We do not believe that one can expect to find a consistent pattern of long term scale economies here. The variation in technology is phenomenal, from the treadle pump up to major infrastructures for river diversion. An extensive review of the use of the treadle pump in Africa shows how much more effective it is than traditional rope-and-bucket irrigation systems. In Bangladesh more than 500,000 of these low-cost devices are in daily use. Even where scale economies exist for one component of the irrigation cycle, such as a river barrage like that on the Bhavani River, distribution costs may be greater across the large area in command.

With the exception of springs and of surface-water diversion through civil engineering structures large or small, virtually all groundwater and surface-water abstraction is carried out by pumping. So the possible existence of long term scale economies in pumps and their power units needs to be reviewed.

Positive displacement pumps such as the piston pump, the rotary pump, the air-lift pump and the Archimedean screw are convenient for small discharges. Of the roto-dynamic pumps (centrifugal, axial and mixed flow) centrifugal pumps are the most widely used. They are ideally suited to small discharges at high pressures but in fact are very versatile. Axial pumps are very efficient at lifting large volumes of water at low pressure; they tend to be used for large pumping works because of the expense of their drive shaft and bearings. For some middle-range head and discharge requirements, mixed flow pumps (combining the centrifugal and axial flow principles) perform best. Centrifugal pumps can also be operated in series when extra head is required.

Pumps and their power units convert fuel energy such as diesel oil into useful water energy. One of the interests of hydraulics is how well the power from the engine or motor is converted into useful water power in the pump. This is given as a measure known as power efficiency. In the absence of friction, efficiency would be 100 per cent but there are always friction losses in all the components of both the pump and its engine. Economies of scale *do* exist here. As Melvyn Kay writes (1998:230): "Smaller pumps are usually less efficient than larger ones because there is more friction to overcome relative to their size."

Force-mode in contrast with suction-mode pump technologies are more expensive but give access to groundwater at greater depth. Here, in particular, the larger investment leads purchasers to give greater consideration to pump and engine performance, reliability, repair facilities and spare parts availability.

With respect to power units, electric motors are very efficient in energy use and can be employed to drive all sizes and types of pump. Where access to electricity is not easy or is unreliable, petrol or diesel engines are used. Diesel engines are heavier and more expensive to purchase but they are more robust than petrol engines, have a much longer working life and are cheaper to run.

Turning now to storage head works, strictly in terms of their engineering economics, dams certainly offer economies of scale. For example, large dams in terms of storage capacity on balance have a greater depth of water and therefore lose less of their water to evaporation. Again, as Melvyn Kay points out in his excellent introduction to hydraulics, the thickness and therefore the cost of a dam wall's construction is determined by the head of water it contains, not by the volume of water it stores. This is known as the dam paradox. In general, the ratio of the number of cubic metres of water stored to each cubic metre of wall increases with dam size. The empirical relation between dam size and capital cost for each thousand cubic metres stored strongly confirms the existence of scale economies.

The costs of storage are also strongly affected by site conditions: in broad valleys with poor foundation conditions, costs will be high. An ideal site is a narrow gorge broadening to a large valley up-stream with good, solid rock conditions and no risk of earthquake activity. In the case of storage, as with other infrastructural works, the best sites tend to be developed first.

Max Neutze stresses that unambiguous scale economies exist for the *networks* of land drainage, flood control and fresh and wastewater services, including irrigation service supply. The cost of pipes increases roughly in proportion to their length and diameter, but their capacity increases in proportion to their diameter to the power of about 2.6. This is because the cross-sectional area increases as the square of the diameter and because friction between the water and the pipe decreases with size. Indivisibility, Neutze argues, is a characteristic of networks.

On the subject of networks, an advantage of boreholes over surface-water surfaces is that they can be sited adjacent to the areas to be irrigated, thus reducing distribution overheads. Stressed that the long-term cost function deals with additions to capacity "in a single leap", that evolutionary development is not in question here. However, when engineers are preparing plans many years ahead, it may be necessary to plot an evolutionary path. In the case of a 10-year water strategy, for example, it may make sense to consider two rounds of investment.

Here one is concerned not with a once-for-all heave to arrive at a given output, as with the long-term supply function, but with a first investment commitment followed by a second round of construction, for example, as an agricultural area's demand expands over time. Average overhead cost for the first of the two investment rounds in our example. But the AOC of the second round is not so defined, since the long term cost function will have shifted as a result of the prior introduction of the first-round infrastructures.

Here is an example. Suppose that the stock of water in a reservoir for an irrigation district when it is fully developed will need to be 100 units. The single-round solution might simply be to build a 100-unit facility. A two-round solution might be to construct a 75-unit reservoir initially and then to add 25 units more in a second stage. The single-throw approach may have the advantage of providing a more cost-effective solution when the district is fully in production but at the cost of underutilized capacity during the first phase of the district's development.

But note, too, that the second-round addition of 25 units might be achieved either by a free-standing 25-unit reservoir or by deepening the 75-unit facility. With a double-throw, the first investment round changes the options available when the second round arrives. So, we can define the *indivisibility* of water infrastructures as what makes a single 20 m dam cheaper than two dams each of 10 m–a cross-section comparison. The *lumpiness* of water infrastructures is what makes a 10 m dam built now impose a higher construction cost for an additional 10 m capacity a year ater–a time-series comparison. Lumpiness particularly characterizes the irrigation networks of pipes and canals. Lumpiness occurs because land prices will have been pushed up by first-round infrastructural construction, because pipe and canal installation is more expensive when it necessitates digging up an existing distributional infrastructure and because second-round investment is more disruptive than that on virgin sites.

If the range of second-round options are part of a long–term infrastructural plan in designing the first-round facility, the 25-unit expansion is likely to be achieved more cost-effectively than without such planning. This is the basis of Neutze's argument that the durability, specialization, immobility, indivisibility and lumpiness of infrastructural provision provide a strong economic argument for land-use planning and the demand management of growth. Note that groundwater abstraction is far less lumpy than surface-water abstraction. Tubewell construction can be phased in with demand, unlike canal systems and dams. The attractions of short-term economies of capacity utilization contrasted with the potential advantages of long term planning and the link between prime and overhead costs, are illustrated by Kay:

> Pipe sizes are often selected using a discharge based on present water demands and little thought is given to how this might change in the future. Also there is always a temptation to select small pipes to satisfy current demand simply because they are less expensive than larger ones. These two factors can lead to trouble in the future. If demand increases and higher discharges are required from the same pipe, the energy losses will rise sharply and so a lot more energy is needed to run the system.
>
> As an example, a 200 mm diameter pumped pipeline, 500 m long supplies a (discharge) of 50 litre/s. Several years later the demand doubles to 100 litres/s. This increases the velocity in the pipe from 1.67 to 3.34 m/s (i.e., it doubles) and the energy loss rises from 5 m to 20 m (i.e., a four-fold increase). This increase in head loss plus the extra flow means that eight times more energy is needed to operate the system and extra pumps may be needed to provide the extra power required. A little extra thought at the planning stage and a little more investment at the beginning could save a lot of extra pumping costs later.

Note that Kay does not envisage adding a second 200 mm pipeline. The correct choice trading off short-term economies of utilization against lumpiness and long-term scale economies needs project evaluation over the life of the scheme. Infrastructural efficiency is not restricted to questions of scale economies. At the two extremes, buried asbestos pipes cost twenty-three times as much as high-density polyethylene (HDPE) surface piping-the distributional choice at Keveral Farm. There is little or no correlation between capital cost and either maintenance cost or distributional efficiency.

But the capital and maintenance costs of conveyance systems between the point of supply appropriation and the irrigated fields are not the only determinants of farmers' choices. Other factors of great importance are the *availability* of a distribution technology at the time of infrastructural construction, the *credit* provided for alternative technologies, the *permeability* of local soils and the hydraulic *head* available for water distribution, van Bentum and Smout (1994), for example, point out that soils that are sandy or loamy favour buried pipe systems. Scarce supply combined with low hydraulic head favours lined channels.

Compacted earth channels score best when soils are neither sandy nor loamy, where water is not scarce and where low head disfavours buried or partially buried pipe systems. In hilly areas the cost of channels is increased because their alignment must follow the land's contours to create a gentle downward slope for the flow. A more direct but steeper route would bring higher flow velocity causing erosion and serious damage to the channels. Pipes can be used in any kind of terrain, can take a direct route and the resulting high water velocity brings no risk of soil erosion. Channels are particularly attractive options for conveying large quantities of water over relatively flat land such as in large irrigation systems on river floodplains.

The Capital Financing of Irrigation Infrastructure

The infrastructural supply of irrigation services requires a wide variety of real capital resources. So the actors responsible for service supply must find the capital finance for these resources' capital costs. John Briscoe has pointed out how neglected is the subject of irrigation's capital financing. He writes:

> To illustrate just how complete this neglect of financing is, a recent major review of the World Bank experience in irrigation...has this and not a word more, to say about irrigation financing: "There are no reliable statistics on global irrigation investment." The subject is of so little interest (including, presumably to reviewers and the audience in the irrigation community) that the subject is not mentioned again in this 140-page report. Similarly, an irrigation policy review for the UK Department for International Development does not mention the issue of financing save for a report on official development assistance flows to the irrigation sector....

The two reports to which he refers were published by the World Bank. In this "data desert" Briscoe himself estimates the aggregate level of irrigation investment in the developing countries. Using a variety of sources, including country-level data published in 1993-96 from Bangladesh, Colombia, Egypt, Ethiopia, India, Laos, Mexico, Morocco, Nigeria and Vietnam, he arrives at a figure of US$25 billion per year. For water-related infrastructure as a whole, including irrigation, hydropower, water supply and sanitation, the mix of funding is 90 per cent of investment from domestic sources and 10 per cent from external sources.

The World Commission on Dams suggests that the annual investment in large dams for irrigation in the 1990s, excluding their distribution infrastructure, was US$8-11 billion in the developing countries. In the developed countries, investment in large dams for irrigation, water supply and flood control was US$3-5 billion. The WCD used the term "large dams" to mean a dam with a height of 15 m or more from the foundation as well as dams 5-15 m in height with a reservoir volume of more than 3 mcm. There are 45,000 such dams around the world.

Historically the actors responsible for service supply have fallen into three groups: farming families financing their own infrastructural supply; agribusinesses in a similar position; and public-sector irrigation and drainage authorities. Broadly speaking the first of these has the smallest capital needs per actor; the third has the largest; and agribusinesses lie in the middle. More recently, non-governmental local entities for collective action, such as water user associations, have paid for machinery and equipment out of their irrigation fee income.

The variety of ways in which capital financing is accessed is great, both within and between these three groups. The same applies to the terms under which finance is made available, such as the maximum loan available, the number of years over which it must be repaid, the rate of interest charged and the security demanded in case of failure to meet the agreed repayments schedule. These terms are best understood in the light of each country's banking and credit system or, in the case of many public-sector irrigation and drainage agencies, with reference to the policies and practices of the international banks. Large agribusinesses may access finance not merely through credit but also through capital raised by issuing shares. Briscoe stresses how important is the supply of irrigation water from the private sector. He suggests that it is as high as 70 per cent in Pakistan and 40 per cent in the Philippines and India. He also associates the form of supply appropriation with the institutional source. "Surface irrigation is the classic government-driven public works and rent-seeking operation .... Groundwater irrigation is usually "out of sight" in the informal, small-scale private sector."

David Sutherland and Peter Howsam have prepared a reference manual on groundwater irrigation technology management that can be used to illustrate investment financing in the case of farming families in Bangladesh and Pakistan. The sources of money for the purchase of pumps and their power units are indeed diverse and include: the farming family's own savings, money borrowed from relatives and friends, the hiring of irrigation equipment, non-governmental organizations (NGOs) providing loans, credit from private or national agricultural banks and loans or cash subsidies from local and regional government. Better-off farmers who succeed in access to credit may supply their surplus irrigation water to smaller farms; this is particularly the case in Pakistan.

However, there are difficulties associated with the supply of credit in Bangladesh and Pakistan. Reluctance on the side of the farming family to borrow may derive from the loan repayments they will have to make prior to harvest time, the complexity of loan procedures, the bribes required by bank staff to issue a loan, the indebtedness risks created by fluctuations in crop turnover and, more generally, the stark poverty that makes repayment of any loan prohibitive amongst the landless and the smallest landowners. On the side of lenders, unwillingness to lend is caused by bad debts accruing on their profit and loss account exacerbated by government waiver of farmer debts, borrowing requirements that exceed microcredit limits and the transaction costs associated with large numbers of small loans. The inability of farming families to mobilize capital restricts the number of entrants to private tubewell ownership or drives many to buy the cheapest technology available, even when the lower capital costs are more than offset by higher costs of operation and maintenance.

The WCD suggests that the first and single largest financier of large dams has often been the World Bank (WCD 2000:19). With respect to the developing countries we can expand on this. Briscoe believes that 10 per cent of the financing of water-related infrastructure in those countries comes from external sources. Of this about one-half is provided by the Bank alone. In 1997 the Bank's portfolio of irrigation and drainage projects had commitments valued at about US$7 billion, about 6 per cent of its total portfolio. These commitments are to the third category of irrigation suppliers referred to above-the public-sector irrigation and drainage authorities.

The World Bank's capacity to lend on this prodigious scale exists only because it also has an ability to borrow of extraordinary magnitude. Most of the Bank's lending is long term; so too is the bulk of its borrowing. This takes place in the world market for bonds. In this 8-month period alone the face value of sterling-denominated bonds and US dollar-denominated bonds issued by the Bank was £765 million and US$4,050 million respectively. Of fifteen issues, only five had a maturity date less than 5 years after issue. Three issues mature in June 2032; where will the reader of this book be then? And where the writer?

The costs borne by the Bank of the bonds it issues are the fees it pays at the time of their marketing and the rate of interest the Bank has to pay on the money it borrows. It is a record for the London market for the last Wednesday in each month (or thereabouts) from August 2000 to March 2001 of the yield on World Bank bonds with a redemption date in June 2010.

But the actual rate of interest earned on a bond with a face value of, say, US$100 depends on *how much the lender pays for such a bond*. This payment may be more or less than the face value. This price. The bid yield on a bond–the rate of interest actually received for a given bid price-is equal to the coupon multiplied by 100 and divided by the bid price. Where the bid price happens to be 100, the yield is equal to the coupon. The bid price always exceeds 100 and so the bid yield of column 7 is always less than the coupon of 7.125 per cent. Simple! It follows that when the World Bank raises fresh capital on a bond denominated say in US dollars, with a redemption date say in June 2010, it knows that the actual rate of interest it will have to pay on that new money is going to be about equal to the current bid yield on bonds of that type.

The market rate of interest on bonds with a long term date has many determinants. Just one is highlighted here and that is the credit rating of the institution issuing the bond. Credit rating is provided by two major agencies: World Bank debt their highest possible rating AAA and Aaa respectively. As a result the market is willing to accept a rate of interest from the Bank on newly issued capital denominated in US dollars only 100 or so basis points above that paid by the safest security in the world, bonds issued by the US Treasury. It ranges from 73 to 116 basis points, i.e., from 0.73 per cent to 1.16 per cent.

With respect to irrigation (and other) projects, the institution that borrows from the Bank is typically national government, not the institution that is responsible for the construction and management of the new infrastructure. An intermediary stands between the Bank and the irrigation authority. As a result, if the project fails the Bank does not suffer financially. There is a break in the line of prudential responsibility for ensuring that the institution requiring loan finance for capital investment is capable of repaying its debt.

Gestation Period Delays and Cost Overruns

The gestation period of an irrigation project is the length of time from the inception of design and construction to completion. The World Commission on Dams has published useful data and analysis of gestation period delays with large dams and their associated investment cost overruns (WCD, 2000:39-42). Dams in the WCD Knowledge Base demonstrate a marked tendency to both delay and overrun with interesting interdependence of the engineering and financing determinants. Note that there are two main categories of large dam. *First,* the reservoir-type storage dam impounds water behind the dam for seasonal, annual and in some cases multi-annual storage and regulation of the river. *Second,* the run-of-river dam often has no storage reservoir and may have limited daily pondage; it creates a hydraulic head in the river to divert some portion of the river flow to a canal or power station.

Large dam *delays* are considered first. In comparing the planned gestation period with actual outcome of the ninety-nine projects in the WCD's cross-check survey, fifty came in on or before schedule, twenty-seven were delayed for 1 or 2 years, eleven for 3 or 4 years and four for 5 or 6 years. Four dams were delayed for more than 10 years. "The existing literature on large dams and related projects confirms this finding: large dams tend to be subject to significant schedule slippages" (WCD, 2000:42).

The causes of delay are cited as contractor and construction management inefficiency, unrealistic construction schedules, late delivery to site of essential equipment, labour unrest, legal challenges by affected groups and, finally, difficulties in raising capital finance for the engineering work, as with the Tucurui Dam in Brazil. The impact of gestation period delay is to postpone both the delivery of the dam's services as well as the inception of the stream of revenues accruing to the dam owner. Economic and financial performance is consequently weakened. Delay also brings with it an increase in the interest charges on the capital borrowed over the gestation period. In the case of Tucurui, for example, more than one-third of its 77 per cent cost overrun is attributable to additional interest payments during construction owing to construction delay.

The general issue of *cost overruns* is now considered. The variability of outcome is high, particularly with multi-purpose projects. There is no clear correlation with dam size. Dams built in India perform worst according to the WCD report. But the overall results are less dramatic when corrected for inflation during the gestation period, falling from an average exceeding 50 per cent at out-turn prices down to 21 per cent in 1998 US dollars.

The causes of investment cost overrun include poor development of cost estimates, technical problems arising during construction, weak implementation by contractors and poor prediction of inflation. The WCD comments:

> Part of the difficulty in developing accurate projections for construction costs of large dams is that the geotechnical conditions at a site (the quality of the rock for the foundations of the major structure and for tunnels) and the quality of the construction materials cannot be determined precisely until construction is under way. Discovery during construction of less favourable site conditions than those assumed in the engineering designs and construction plans can be a significant contributor to cost overruns and delays in time schedules. Despite being a common factor in causing overruns, little to no provisions have been made to improve the estimates in this regard.

The WCD suggests its evidence confirms that there is a systematic bias towards underestimation of large dams' capital costs.

Dam projects are usually given the go-ahead on the basis of a financial budget for the investment. The impact of substantial overruns is that additional funds have to be found, sometimes creating difficulties for public or private budgeting. Where tariffs are based on cost estimates, overruns can undermine cost recovery and the project's financial rate of return.

CASE STUDY: EUROPEAN INVESTMENT BANK ACTIVITY IN THE MEDITERRANEAN

This case study of project financing by the European Investment Bank (EIB) in the Mediterranean takes up the issue of prudential responsibility. The EIB was set up in 1958 by the Treaty of Rome, is owned by the fifteen European member states and has its headquarters in Luxembourg. It supports European Union (EU) policies on a self-financing basis, raising its resources on the world's capital markets for onlending, it says, to sound capital investment projects that promote the balanced development of the EU. As a major international borrower that has always been awarded the highest AAA credit rating by the world's leading rating agencies, the Bank raises large volumes of funds on excellent terms. It onlends the proceeds of its borrowings on a non-profit basis. While the bulk of its loans are within the European Union, the Bank has also been called upon to participate in the implementation of the Union's development aid and cooperation policies in some 120 non-EU countries.

The EIB's financing activities in the EU's member countries in the Mediterranean began with the Bank's inception. Financing in non-member Mediterranean countries began in 1962 and has chiefly taken the form of long term loans, although money has also been made available as risk capital funding designed to encourage development of the local private sector and in joint ventures.

The countries forming the boundary of the Mediterranean Sea and its island states share a broadly similar climate. The summers are dry and hot and the rainfall pattern can result either in severe floods or in droughts. Precipitation varies from an annual average of 1,000 mm in the mountainous areas of the north to less than 100 mm in the south. A wide diversity in the per capita water resource exists; it is generous in some countries of the Balkan sub-region but arid in countries such as Malta and Tunisia. Water use per capita also varies considerably, from 100 m^3/ year at the low end of the range to more than 1,000 m^3/year in countries with a large irrigated sector.

It is the southern and eastern countries in the region that most experience water stress. This has led to the mining of deep aquifers, sea water desalination and the use of land-sourced brackish water at high unit costs. "The situation is particularly serious where the environment is under severe threat, where expensive water is wasted through distribution pipe leakages, rationed by intermittently cutting the supply to consumers, or wasted through inefficient irrigation systems and practices." The Bank believes that the reuse of treated waste water for irrigation and industrial purposes has an important part to play in the development of integrated solutions for water resource and environmental pollution problems.

In 1999 the EIB published a review by its Evaluation Unit and by outside consultants of seventeen water projects it had financed around the Mediterranean Sea mainly during the 1980s. This case study is based on that report. The projects selected are located in eight countries north and south of the Mediterranean, they accounted for 11 per cent of the projects financed by the Bank from 1981 to 1992 and they represented a total investment cost of 1.4 billion European Currency Units (ECU) and total EIB loans of 430 million ECU. Forty per cent of the investment went for five irrigation projects and the remainder to water resources development, water distribution, sewerage and wastewater treatment. Projects selected were drawn from both EU member countries and those outside the Union. Water projects within the EU are usually financed with the objective of fostering regional development and/or environmental protection. Projects outside the EU are financed under mandates given to the EIB by the European Council in order to promote economic development.

The investment total of 1.4 billion ECU included land acquisition, engineering and supervision, civil works, supply of equipment, working capital and interest during construction. The irrigation projects included extension services and these projects' cultivated areas ranged from a minimum of 200 ha to a maximum of 65,000 ha with a median value of 6,000 ha. Schul writes:

> Three out of the 5 irrigation projects were implemented according to the original design capacity. One particularly risky project seeking to introduce drip irrigation to stop over-exploiting a non-renewable water source failed entirely, despite intensive EIB technical assistance. In the end, agricultural rather than irrigation equipment was purchased to use up the EIB loan. The technology proposed was over-sophisticated for the local management and the EIB became aware of corruption after the loan was fully disbursed. The institutional set-up has since been changed. Of the seventeen projects, only three were completed on time; the implementation delay on the others ranged from 2 to 12 years. The problem was particularly acute in two EU countries.

Schul further comments:

> Unreasonably slow implementation resulting from institutional weaknesses is one of the main curses of water projects in the Mediterranean area. This has been one of the factors contributing to the EIB's decision to no longer finance irrigation projects in some countries. In one symptomatic case, for instance, the EIB participated with a local bank in an innovative deal where farmers agreed to finance 65% of the investment costs of collective irrigation schemes. Farmers expected this to solve the problem that 100% subsidised schemes were being continually postponed because of their Government's budgetary restrictions. However, after a few years they withdrew from the arrangement, partly because the Government interfered with procurement procedures, which resulted in delaying access to irrigation water by sometimes several cropping seasons. In the meantime the Government managed to accelerate its own programme with EU grants providing 100% subsidies.

Six of the seventeen projects experienced difficulties common to public procurement procedures. These were:

- Excessive delays between bid receipt and contract award.
- Bids accepted from inexperienced contractors or extremely low-cost tenders forcing subsequent renegotiations, quality problems and implementation delays.
- Changes in design, disputes with suppliers and the bankruptcy of appointed contractors.

Investment costs per hectare of irrigation works co-financed with the ultimate beneficiaries were 50 per cent or less of scheme costs fully financed by state authorities. In four of the five irrigation projects, production was equal to or greater than that originally forecast. The fifth project produced nothing. Two projects were limited to smaller irrigated areas than first planned because rapid urbanization reduced both the available supply of water and the farm hectarage. Three of the five projects made possible a reduction in desertification or soil salinization or soil erosion on steep slopes. Treated waste water was reused in two of the projects for irrigation supply. The loan characteristics of the seventeen projects deserve consideration. Project investment cost, the rate of interest, the grace period and the repayment period are variable to a staggering degree. The huge gap between the minimum and median values of the loan rate of interest deserves explanation. The EIB's own policy on external lending is to charge a standard mark-up: 25 basis points above the cost of borrowing. But the EU budget provides grant funds for a number of the mandates, including the Mediterranean, part of which is available to subsidize the interest rate on EIB loans.

The EIB states that it lends money on "sound capital investment projects". What is equally startling is that the median *ex post* financial return is negative. Finally the median *ex post* economic return is a full 320 basis points below the *ex ante* calculations. Low financial rates of return are attributable variously to the cumbersome decision-making procedures of the public sector, excessive operating costs, low charges for water, the disbursing of loans at loan signature and excessive implementation delays. In only two out of the seventeen projects did financial profitability match EIB lending rates.

The Review's summary points out that most water resource projects and agencies are financially unprofitable, with irrigation projects not performing significantly worse than the other sub-sectors. In respect of irrigation projects within the EU farmers deviated from the original cropping patterns and adopted crops subsidized by the Common Agricultural Policy. Here farmers pursued financially rewarding activities, but ones that, from the point of view of the Union as a whole, had negative economic returns. We have noted that the EIB is a self-financing institution. How then is it feasible for the Bank:

- to raise money in the international bond market on strictly commercial terms;
- to on-lend these same funds at interest rates that cover both the Bank's own debt repayment obligations as well as its costs of operations;
- over an 11-year period to appraise and approve loans on seventeen water resource projects that in retrospect enjoyed financial profitability that matched the Bank's lending rates in only two cases and where the median *ex post* financial rate of return was negative?

To answer this question we must return to the issue of prudential responsibility and its fragmentation.

The institutional category of both the borrowers of the EIB's loans and the projects' promoters. Apparently all these institutions were in the public sector, even the Société Anonyme. Only two loans were granted directly to the project promoter; the remainder went first to an intermediary that channelled funds on to the promoter. In thirteen cases the borrower was a central government ministry or a state bank.

Typically, the full loan is disbursed at loan signature or in tranches with a view to matching expenditure. This is particularly the case within the European Union, where the EIB's loans are made for projects in an advanced state of implementation. The Bank's records show that, after the initial project appraisal, EIB staff visited water projects in the Mediterranean *only once in every 5.5 years*. Loan provision to government or some other intermediary usually means that the financial terms on which the loan is made to the middle-man is quite different from the terms between that middle-man and the beneficiary of the loan.

The fragmented relationship between the Bank and the final recipient of a loan exists, then, because loans are typically made through an intermediary, the loan is disbursed on signature and EIB follow-up is weak. But there is a fourth, complex issue that arises from the channelling of Bank loans through government (or any other) intermediary. This is known as the "fungibility" issue and it requires explanation.

Imagine that a multilateral finance institution (MFI) such as the EIB or the World Bank lends money to an irrigation authority for construction of new storage requirements. The MFI stains the notes red and delivers them directly to the authority (there is no intermediary) and insists in the financial agreement (which is honoured) that all these red notes and only these red notes are spent on the new infrastructure. In this unusual case we have no doubt whatsoever that the loan does indeed finance the project.

But the world does not spin like this, as we have seen. What actually happens is that the MFIs loan goes to an intermediary such as a government department that probably places the money in an interest-earning account where it joins financial inflows from a multiplicity of other sources. When the irrigation authority draws down the loan made to it for dam construction, there is no direct link between the financial transfer from department to irrigation authority and the original loan from the MFI to the department. Indeed it may well be the case that the department-to-irrigation authority loan would have gone ahead even in the absence of the MFI-to-department loan. In such cases we speak of the MFI loan as meeting the "treasury requirements" of the department. The non-existence of a direct link between the money loaned by the MFI and the money borrowed by the irrigation authority is an instance of what is called "fungibility".

Here is a hydrosocial parallel. Three streams feed a reservoir. The waters from the dam are channelled along a river past seven farms that divert the flow for crop irrigation. There is no sense in saying that any single stream sources any single farm. The *Concise Oxford Dictionary* says of the adjective "fungible" that it is a term in law, with reference to goods contracted for, when an individual specimen is not meant. In this context "fungible" means "that can serve for, or be replaced by, another answering to the same definition".

The central conundrum of fungibility is as follows. However tightly an MFI ties its financing to a specific project, a national government will have a wide range of projects for which it is seeking finance. If the MFI provides capital finance for a high-priority project that government would have managed to finance anyway, then the MFI loan permits government to transfer the funding released to the highest priority for which funds are *not* available, one which the MFI may or may not approve of. However, if the MFI provides capital finance for a low-rated project, truly additional funding in this case, that project may well not be one which the MFI wishes to see go ahead. As Tony Faint writes (personal communication), this is a dilemma for all project-based lending and the EIB is essentially a project-financing institution. Schul's (1999) excellent report makes all of this clear. He recognizes the uncertainty in the meaning of the Bank's "contribution to objectives through its investments" when projects would have gone ahead even in the absence of Bank financing.

Whilst the Bank essentially makes loans to sound, creditworthy borrowers, this may have contributed to weakening the link between EIB loans and the projects. Loans were usually channelled through intermediaries and resources were then allocated to the final beneficiary on the intermediary's own terms, rather than on the EIB's. Hence, lending terms were either not well adapted to the investment or the promoter, or EIB funds *de facto* financed the treasury requirements of the intermediary institution or public administration. This may have reduced the EIB's financial exposure, but its scope to influence the outcome of projects and impose the discipline that usually accompanies bank lending was also diminished.

The Report's conclusions include the view that the link between its loans and the targeted investments needs to be tightened in order for the Bank to be able to apply appropriate loan conditions and bank discipline on promoters. Loans should be channelled as much as possible to the promoter. In addition, projects should be monitored at least once per year with regular reports to management on specific portfolios.

PRUDENTIAL RESPONSIBILITY: BREAKING THE LINE

The point that irrigation's capital finance is a neglected subject. In particular the powerful role of the World Bank is discussed. It is an attractive financing source for would-be borrowers because of the magnitude of its lending, the long redemption dates of its loans and, in relative terms, the attractive rate of interest that it demands on those loans. The Bank's portfolio of commitments to the irrigation and drainage sector in 1997 was some US$7 billion. But these are *not* loans made to the irrigation authorities but to government. As a result, if a project fails the Bank does not suffer financially. There is a break in the line of prudential responsibility for ensuring that the irrigation authority receiving loan finance is capable of repaying its debt.

EIB loans totalling 430 million ECU were provided for seventeen projects. In fifteen cases the loan went first to an intermediary, usually a central government ministry. There is no way, at least in these cases, that we are convinced by the EIB's claim that its resources go to "sound capital investment projects".

A disturbing general scenario arises for all cases where the line of prudential responsibility is broken in this way. An MFI may embody institutional routines that promote project identification and loan approval where a cool appraisal of the project's strengths comes second to the desire of the MFI to build up its loan portfolio. In such cases a pattern of lending would be established where projects that cannot be objectively justified go ahead in any case but where the MFI is protected from the financial consequences by the intermediary institution that is the recipient of the loan.

Chapter 7

Operation, Maintenance and Management

FIELD SUPPLY EFFICIENCY AND FULL COST PAYMENT

The tasks of operation, maintenance and management for irrigation service supply have already been described a number of times in this book, particularly in the case studies of the Lower Bhavani project, the Curu Valley and Keveral Farm. These tasks consist of running the infrastructural supply headworks and networks between the locations at which irrigation water is socially appropriated–the five base components of supply-through to the points at which the water reaches farmers' fields.

Such tasks include OM&M of pumps and their power units, monitoring and repairing water storage investments, desilting and clearing weed infestation from distribution channels and irrigation scheduling in the supply of water to farmers. Where restorative maintenance or asset replacement is required, financial outlays may be classified as current or as capital costs, depending on the scale of the work.

The distinction has been drawn between prime and overhead costs and the existence of short-term economies of capacity utilization in average *overhead* costs was suggested. Such economies also exist for *prime* costs. Average prime costs are lower at higher rates of capacity utilization of a given infrastructure primarily because workforce costs in appropriation, storage and distribution increase only modestly at higher irrigation water volumes. Workforce costs are "sticky" in respect of the OM&M of a given irrigation capacity.

There may also exist long-term scale economies in prime costs. Costs refer to the appropriation, storage and distribution of the base supply up to the farmer's field. The volume S_b is the base flow appropriated, measured for example in cubic metres; the volume S_f is the flow in cubic metres that actually reaches farmers' fields after drainage and evaporation losses.

Prime costs plus overhead costs equal total costs. Both average overhead cost and average prime cost decrease as capacity utilization increases and so average total cost is also lower. We now have:

$$E_8 = \frac{K}{S_f}$$

where E_8 is *field supply efficiency*, K is the total costs of the base flow's appropriation, storage and distribution and S_f is the field supply. Note that by definition the *total cost* of the base supply and the total cost of the field supply are the same. But the *volume* of the field supply always falls short of the base supply volume.

So the *average cost* of the field supply always exceeds that of the base supply because of the difference in these two volumes. Carruthers and Clark point to advantages that groundwater abstraction has over surface-water abstraction in respect of field supply efficiency:

> Groundwater storage is not subject to significant evaporation losses and because the water is used close to the well, conveyance losses are reduced to a minimum. Wells have flexibility in that they can be operated at any time according to farm demand.

The main disadvantage of tubewells is that they generally have high recurrent costs. If the wells are powered by electricity, these costs can be minimized by using as much surplus or off-peak power as possible. This will often result in night-time irrigation, in which case application efficiency could be maximized because evaporation losses are lower at night. However, in some communities there are strong reactions to night-time irrigation, which lead to poor field supervision. In addition, the lack of ability to see may lead to wastage rates greater than the savings from lower evaporation.

With the concepts of overhead, prime and total cost all now in place, the relation of these costs to the cost of water met by the farming family or agribusiness. *A cost to the farmer per cubic metre of field supply that is equal to the average total cost of irrigation water we shall call full-cost payment.* Full-cost payment is feasible in the cases of own-supply cost, a volumetric price, an indirect fee and a revenue-only fee. But in the case of a zero charge, or other cases where the cost to the farmer falls short of the full-cost payment, something has to give.

It may be that loan payments are not made, or public subsidy is required, or maintenance by the irrigation authority is neglected. In the last case it is likely that the capacity utilization of the irrigation infrastructure will fall–resulting in an even higher average total cost as the short-term economies of capacity utilization are lost. A downward spiral is in place as equipment malfunctions multiply, distribution losses mount and silt and weeds degrade the system's hydraulic efficiency.

THE VALUE OF IRRIGATION SYSTEM MAINTENANCE

John Skutsch and Darren Evans (1999) have made a valuable

contribution to the debate on irrigation maintenance and this section draws heavily on their work. They suggest that widespread evidence points to a considerable shortfall between recommended maintenance expenditure for public-sector irrigation schemes and the amounts actually spent. In a variety of projects researched during the 1990s in India, Indonesia, Pakistan and Sri Lanka, the ratio per hectare of actual spend to recommended spend was only 24-47 per cent.

In the first years of a new scheme, maintenance neglect may have no dramatic effect. Irrigation staff can freeride on the freeboard produced by the vertical distance by which the channel flow's usual height exceeds the required operating level. Cropping systems, too, may show some resilience. But sooner or later maintenance underfunding begins to bite. Schemes in arid and semi-arid areas, where crops are entirely dependent on irrigation under managed rotations, are more sensitive to neglect than projects in the humid tropics that have paddy rice under continuous irrigation supplementary to monsoon rains.

The biggest losses come through crop output decline as the volume of irrigation water delivered eventually begins to decline. This may show itself as a fall in tonnage efficiency or in a diminution of the irrigated area. Moreover, as Carruthers and Morrison indicate, farmers shift to lower value crops so as to be able to reduce risk by limiting the use of inputs. Impeded drainage can lead to waterlogging and salinity and to an increase in the incidence of water-related diseases associated with blocked channels and stagnant water. Skutsch and Evans, commenting on the serious social and financial implications for farmers, write:

> Those in the more favoured parts of the system may continue to receive an adequate water supply, while those at the tail-end can face ruin. Poor maintenance thus directly aggravates existing inequities within the farming community. It initiates a vicious circle of decline: reduced water supply; lost output; farmers' anger, despair and reduced investment; reduced water fee collections; vandalism and conflict. Smallholder farmers operate on a narrow margin between relative success and failure. Land is often mortgaged against the following harvest to pay for the cost of agricultural inputs. A single disastrous cropping season can mean the loss of a farmer's land and enforced migration to the cities to seek work.

This process brings with it the premature obsolescence of irrigation projects. The search for new capital funding begins in order to renovate the assets far earlier than would have been necessary under a satisfactory maintenance regime. A cycle of build-neglect-rebuild is in place. Partly as a result of this, most development bank finance for irrigation is now for the rehabilitation of existing schemes. Skutsch estimates that about two-thirds of recent international lending has been for systems that have suffered early technical failure. Already in 1995 Jones's overview of World Bank lending for irrigation projects could conclude: "O&M problems can be seen in the Bank's financing of so many rehabilitation projects. Almost all of them, when scrutinized, turn out to be deferred maintenance projects. Why, then, is maintenance neglected? As so often is the case in the social sciences, the question is simple but the answer is fiendishly complex. In part, farmers themselves are responsible. As is commonly the case with public-sector irrigation schemes, they see maintenance as the proper task of government".

If farmers are required to allocate greater resources to maintenance, either in terms of increased payments or of increased responsibility...they must see a clear incentive in better cash returns, reduced costs or improved control over water. Particularly on rice-growing schemes in monsoon climates, farmers may see little urgency in paying today to avoid increasing problems tomorrow, particularly if the government has traditionally stepped in to rehabilitate defective systems. In part, government and irrigation department staff are responsible for the neglect of maintenance. Too often, national, regional and local élites regard the public sector as a milch cow whose teats are there to be squeezed for easy profits. No simpler way exists of doing this than by pushing superfluous staff into administrative posts as a favour to family, professional or political allies. As a consequence the "establishment" of the department becomes excessively large, draining resources from OM&M. Moreover, operations activities in any case enjoy precedence over maintenance. Technical and engineering staff, trained in design and construction, are likely to be unmotivated by the bread-and-butter tasks of OM&M, which offer poor rewards and prospects. For senior staff, administrators and engineers, new projects are likely to be the most highly sought after given their prestige and the rent-skimming opportunities they offer. A rule of thumb of some international consultants in parts of the developing world is that capital rather than current spending is favoured by country élites in government because 40 per cent of capital costs go in pay-offs.

At a level more mundane but of real importance, irrigation departments commonly remit the charges they collect from farmers to the finance ministry. That ministry may give other sectors more support than agriculture when it provides revenue financing. As the sum allocated to OM&M is decoupled from the fees collected from farmers, irrigation agencies have no incentive to maximize their revenue collection. There is a parallel here with the breaking of the line in prudential responsibility, itself a "decoupling" of loan provider and loan user. In part, the development banks themselves are responsible for maintenance neglect. If they lend money on schemes that 15 years later are choked with weeds, silted up, waterlogged, saline, exhibiting high distribution losses and with defective equipment, then their original cost-benefit analyses must have been *seriously* erroneous. The McNamara effect and a weak record in post-project evaluation are partially to blame here. The development banks need not fear that their loans will never be repaid, because the dollar flow of principal and interest comes from national government treasuries, not from the deficient project. Several authors have proposed policy changes aimed at ending the build-neglect-rebuild syndrome. Such proposals include the following:

- An information system should be put in place that is capable of defining in detail the volume of water used.
- A dependable delivery system should be achieved.
- An agency willing and able to collect fees should be built, with a transparent management system.
- Farmers should be required to contribute an agreed number of maintenance days each year.
- Fees paid by farmers should increase but only after:
  - the managing agency details all costs, levels of service and benefits,
  - users are surveyed on their willingness and ability to pay,
  - the charging mechanism is agreed,
  - users participate in setting the levels of service and
  - fees are linked to the level of service.
- Recognized sanctions for non-payment should be enforced.
- Feeding back the charges paid by farmers directly to OM&M.
- Maintenance should be carried out by private or autonomous self-accounting agencies.
- Engineers and planners should design irrigation projects for low maintenance, less costly management and greater effective lives.
- Improvement and modernization projects would be considered only for those schemes where a regular review of their individual maintenance standard shows them to be satisfactory or better.
- Lending agencies should provide financial support for a transitional period following construction.
- Tradeable abstraction rights should be introduced.
- Irrigation management should be transferred to farmers themselves.

MANAGING COMMON PROPERTY RESOURCES

The above list of thirteen proposals for ending the build-neglect-rebuild cycle ends with irrigation management transfer from the public sector to farmers themselves. In the 1990s of such a possibility was able to draw upon and contribute to, a vigorous literature exploring the subject of common property resource management. The published papers and books demonstrate a welcome (if rare) willingness of economists and sociologists to work together in common cause. Below, synthesis of the relevant literature, particularly as it applies to the management of irrigation systems. Major sources are Veblen, Coase, Commons, Hardin, Myrdal, Williamson and North.

In every society common property resources exist, each of which is the common location of work by human agents or actors–the two terms are deployed here interchangeably. Agents may be individual persons, families or institutions. Such common property resources may be constituted by specific natural environments, such as a commons used for grazing livestock, or by some complex resource integrating features of the natural environment with the means of production created by human society, such as an irrigation scheme. The "common property" characteristic of the resource is that a number of separate actors enjoy rights of access and use of the resource-rights that are recognized in law or are customary practice.

In the day-to-day work located on the common resource, each agent may be motivated only by the *private* interest of that person, family or institution. This is of special importance in three ways. *First,* each actor may seek to appropriate as much of the resource for itself, without regard for other agents' interests. *Second,* each actor may choose to maintain the productivity of the resource only to the degree that the private costs borne by the actor in such efforts are outweighed by the private advantages, without regard for other agents' interests. *Third,* insofar as the use of the common resource creates negative side-effects, the actor may seek to reduce these side-effects only to the extent that their cost outweighs their production advantage, without regard for other agents' interests. So, for any single actor, private interest may prevail over the common interest in respect of resource appropriation, resource maintenance and resource degradation.

Production of outputs from a common resource–the milk yield of ruminants or the tonnage of grain-may grow vigorously over time. Work motivated by the private interests of the agent can stimulate long hours in the field, a sharp tactical appreciation of gains and losses from adaptive action and an eager search for innovative practices. In respect of the productivity of the common resource, these are powerful advantages.

However, with the passage of time the common property resource itself and the patterns of productive activity on it may show signs of enfeeblement and impending long-term collapse. Now, the disadvantages imposed by each single agent's actions on other persons, families and institutions have become disproportionately large. The struggle by actors over the appropriation of the resource can weaken friendship, create mistrust and result in disabling and destructive legal and physical conflicts as well as the exhaustion of the resource itself. Maintenance and protection of the common resource may decline below what is required for its long-term sustainability. Each agent's indifference to the negative effects of its actions on others may bring with it both bitter social disputes as well as the poisoning of the common resource.

Such a trajectory may lead to demands for a reconstitution of the social relationships between actors in the management of the common property resource. Where government cannot or will not lead this collective activity, a non-governmental local entity for collective action (LECA) may be created. The core objectives of the LECA are likely to be fourfold. *First,* the shared use of the resource should be recognized by the participant agents as being equitable. *Second,* the maintenance of the resource should be adequate to ensure its long-term viability. *Third,* the production of negative externalities should be sufficiently well regulated that their collective cost to the new institution's actors is acceptable and does not threaten systemic sustainability. *Fourth,* the transaction costs of meeting the first three objectives are acceptable to actors in the light of the benefits they bring.

There is a problem. A successful LECA brings all-round advantages to its actor-members. But every agent knows that if *he and he alone* continues to pursue his private interest, he reaps the benefits of unrestrained action as well as the benefits deriving from the collective agreement. This is known as the freerider problem. As freerider numbers grow, the advantages of collective action diminish and the LECA collapses.

So each LECA needs to engage in forms of moral persuasion in order that its actors honour the collective agreement made. It will also monitor members' activities to ensure that they do not breach the rules and that, when this occurs, sanctions are exercised against such infractions. The collective also faces costs of bargaining, contract formulation and information search in pursuit of a collective economic strategy. Together, all these costs of motivation, control and coordination are termed transaction costs. Even in a study as exhaustive as Anna Blomqvist's in South India it was never possible to assign these costs a monetary value. The greater the LECA's legitimacy, i.e., the greater the internalization of its values and ideas within and outside the institution, the less costly is monitoring and enforcement. Blomqvist concluded from her experience that LECAs are most likely to succeed when the members are culturally homogeneous, small in number, highly dependent on the common resource and benefit equally from collective action.

ABSTRACTION CHARGES

Rivers, lakes and aquifers are common property resources, except where access to them is privatized. This is why common property resource analysis, including the appraisal of transaction costs, is of such great relevance to a full understanding of the management of water resources. In particular, rivers, lakes and aquifers are resources from which water is abstracted for use in agriculture, industry and households. They are also resources used as sinks for the disposal of waste water and drainage water.

Abstraction usually takes place where the abstracting actor enjoys formal legal rights to do so, or where abstraction is an accepted customary practice. In such cases the abstractor may be required to pay a charge either to a public agency or to a private owner of rights in water. Abstraction rights may also be traded.

One objective of these two sections is to provide a general economic analysis of the subject capable of international application to specific case studies. The second objective is to recommend the most appropriate basis for government abstraction tariffs. The scope of abstraction charging includes water consumers abstracting directly for their own use, such as farmers, mining companies and manufacturers. One also includes water service companies that abstract not for their own use but in order to provide a public water supply to irrigators and to urban areas.

A useful analytic starting point is to establish that abstraction charges are a form of economic rent. The modern theory of rent was first developed in the early nineteenth century. Its application was to British agriculture and the common situation where capitalist farmers used land owned by the aristocracy and the gentry and paid rent for that right.

At that time, agriculture contributed about one-third of Britain's total output. The classical school of economists argued that the appropriation of land has the consequent effect of the creation of rent. Thus, Ricardo writes: "Rent is that portion of the produce of the earth, which is paid to the landlord for the use of the original and indestructible powers of the soil". It is true that we may now doubt that any power of the soil is indestructible. Nevertheless, since Ricardo's time, economists have used rent theory whenever they are dealing with a natural resource of economic value that can be privately appropriated and that is in restricted supply. Clearly, this makes rent theory applicable to natural supplies of freshwater. These flows are of economic value, rights in their abstraction can be privately appropriated and groundwater as well as surface water are in restricted supply in any given year.

However, the supposed inelasticity of the water supply deserves closer scrutiny. Within a catchment, short-term elasticity is perversely high where substantial stocks are held in a catchment's reservoirs or where aquifer stocks are high and groundwater pumping capacity is not fully utilized. But it is the elasticity of the planning supply function of water over the medium and long term that is at issue here. It is certainly true that total rainfall in the catchment less evaporation sets a hydrological constraint on long-term catchment abstraction, as Dubourg suggests. But the recycling and reuse of water weakens this natural barrier whilst the desalination of sea water and abstracted water imports from another catchment, can smash it.

The problem is that the wider one casts the net for additional supply, the greater are its unit costs. It is the exponential character of this planning function which maintains the truth of the statement that, in the majority of the world's catchments with substantial populations, abstracted water is indeed in restricted supply.

Returning to the classical theory of differential rent, this assumed that competition existed between farmers for access to land and between landowners in supplying land. In such a situation, it is theorized that land rent on the least fertile tract of land worth cultivating would be zero and here the farmer would earn the going rate of profit on capital for the private sectors of the economy as a whole. Land which was more fertile (or better located) would bring the owner a higher flow of rental receipts such that the farmer, after paying this higher rent, would still receive only the going rate of profit. Rent could therefore be seen as the transformation of the surplus profits of capital, driven by the competition for land of differential fertility.

Ricardo had only a limited interest in non-competitive markets in the supply of land for rent. As a consequence, his theory is essentially demand-driven. This neglect of the supply-side determinants of land rent restricts the scope of differential rent theory in its application to abstraction charges levied by government. Such charges typically exist where an agency of the state licenses abstraction to specified individuals and institutions and where no property right exists to abstract ground and surface water without such a licence. Here, property rights on the supply side are assigned by a state monopoly. So analysis needs a means to interpret the legislative practice of charge-setting by state institutions in different catchments, regions or countries.

Having argued that abstraction charges are a form of economic rent, but that differential rent theory does not provide a basis to explain charge-setting by the state, the best way to proceed is to develop a taxonomy of charge-setting principles. Such principles can then provide the basis for interpretive empirical work in specific catchments, as well as a starting point for policy review of existing legislative practice. Various charge-setting principles are set out below and each is glossed in turn. Note that these charging principles are in respect of the right to appropriate the scarce resource for outstream purposes. They do not include the cost of supply-side infrastructure. Nor are the public-sector costs of flood management, navigation, etc., included here.

- *No charge:* This is the lower limiting case. In the majority of the world's countries, government abstraction charges simply do not exist. In some cases the argument is that, since surface water and groundwater are a gift of Nature or of God, the state has no business in taxing it. This may explain the situation in Scotland, for example. In other cases, what is lacking is the administrative capacity to levy the tax.
- *A revenue-maximizing charge:* This is the upper limiting case. In principle government could raise all its revenue requirements from this single tax. In practice, of course, such a principle is never applied.
- *A market-clearing charge:* In countries and regions that are arid or semi-arid, or where levels of water consumption are high compared with effective rainfall, users may wish to gain access to more water than is available, at least in the absence of inter-catchment transfers. In this case, government may put in place a demand-management policy in which a general abstraction charge is applied that, although it is not revenue-maximizing, does broadly match the demand by abstractors to the annual flows available. No pure examples exist of this, the closest being the public auction of tickets for fixed time and flow in the centuries-old water market of Alicante in Spain. Schiffler and his colleagues at the German Development Institute recorded for the case of Jordan that supply-fix policies in the early 1990s were under considerable pressure and that a demand-management philosophy was taking shape. Politically, it was impossible to tax water abstracted for use by farmers, but an abstraction tax on industry had been introduced there.
- *An environmental regulation charge:* In this case one is considering a country which has a well-developed national policy for water resource management. The necessity for a regulator of the freshwater environment is accepted and abstraction charges are levied and hypothecated to finance the costs of regulation. ("Hypothecation" is a term widely used in economics to refer to cases where government income from a defined tax is reserved

for a specific expenditure category.) This charge may take the form of an average total cost levy equal to the total financial costs of regulation (including compensation payments for rescinded abstraction rights and any other miscellaneous items) divided by the total volume of water abstracted during the year.

- *A Pigovian charge:* Here, the principle is that government should set a charge which is differentiated according to the external costs imposed on society by each class of abstractor. K. William Kapp later formulated the parallel concept of "social cost", covering "all direct and indirect losses sustained by third persons or the general public as a result of unrestrained economic activities". An example of the kinds of damage done by over-abstraction is that of the Hueco Bolsón aquifer on the Mexican-USA border. Over a period of 70 years the water table fell by 25 metres (m), resulting in increased pumping costs, subsidence and contamination by increased flows of saline and polluted waters into the freshwater source. As a general rule, whilst abstraction charges may include a component in recognition of overabstraction, the *differentiation* of the charge on the basis of the monetary evaluation of the external costs never takes place because of the extreme difficulty of measuring them. Indeed, many institutional economists even deny that such measurement has any meaning. Kraemer refers to the overwhelming problems of applying a Pigovian tax in his introduction to the development of German abstraction charging after 1988.
- *An incentives charge:* Incentive charging in a catchment or region can be defined as the use of a water tariff–a table of fixed charges-to give price signals to abstractors that reinforce water resource management based on environmental standards and regulatory controls. The best-developed system of incentive charging is probably that of the Federal Republic of Germany. An incentive tariff can include the following components:
  - A licence fee to have a new water abstraction installation approved by the regulatory authority. For example, Foster in their comprehensive introduction to the hydrogeological, legal and administrative aspects of groundwater licensing in Latin America indicate that installation fees are common and are hypothecated.
  - A charge per unit volume. This may be invariable with total volume consumed. Alternatively, it may rise with the volume drawn off. The total charge payable may be based either on the licensed volume or the actual volume

abstracted. As Schiffler point out, a significant drawback of a rising block tariff is that it hits hardest those abstractors requiring large volumes simply because of the size of the farm or factory. It should also be observed that abstraction charging is never imposed where the installation's capacity falls below a minimum level set by government. This is the case at Keveral Farm.

- A charge reduction for the quantity of water directly recycled to surface waters after use. To take the example of Didcot power station in the UK, in the mid-1990s this had a licence to abstract 142 million litres of water per day. The licence required 50-66 per cent of the water abstracted to be returned to the river, depending on flow conditions. Unit price was lower because of this non-consumptive use. In contrast, spray irrigation provided virtually no return flow to river or aquifer and so was undiscounted.
- A charge that is greater for higher quality water. This is found in Germany for water drawn from deep aquifers.
- A charge which varies with the seasons. The volumetric rate is higher in those months when demand is greater and higher too when precipitation is less.
- A charge which is greater for certain locations. These include:
  (*a*) upstream sources, because the length of the river exposed to abstraction impacts is greater,
  (*b*) rivers, lakes and aquifers most threatened by past or present overdraft, and
  (*c*) regions with lower effective rainfall.

ABSTRACTION CHARGES AND SUSTAINABLE CATCHMENT MANAGEMENT

The subject of abstraction charging is related to the debate on sustainability, beginning with the approach of Richard Dubourg whose contributions to the study of hydroeconomics are within the neoclassical framework. Dubourg suggests that aggregate sustainability is a situation where natural capital as well as non-natural capital are non-declining and specifically where "critical capital" such as water, within the natural capital category, is non-declining.

From these definitions he deduces impeccably that aggregate sustainability is consistent with catchment management policies in which abstraction of surface water and groundwater is equal to effective rainfall. It has to be said that the adoption of such a definition of sustainable abstraction would be extraordinarily dangerous. Abstraction at a rate equal to effective rainfall may be capable of leaving surface and groundwater stocks unchanged over the course of a year so that the critical capital stock is maintained. But a rate of abstraction at this maximum rate would have the effect of capturing the entire river flow at some point or points in the catchment. In effect, Dubourg's definition ignores completely what in the USA is termed the environmental need for water in river systems–all on the basis of three mathematical constraints. In *Economics of Water Resources* the author has suggested an alternative, multi-dimensional approach. After defining the concept of a "sustainable society", it is suggested that water resource planning in such a society has six principal fields of action:

- The protection of water's hydrocyclical capacity to renew its ground and surface-water flows and stocks.
- the conservation of society's species and natural habitats in all their fresh and saltwater environments.
- The husbandry of water in its supply and use.
- The supply of freshwater sufficient to meet the biological, cultural and economic needs of society's human populations.
- The purification of water from domestic, agricultural and industrial effluents.
- The drainage of water and the protection of rural and urban communities against flood.

Abstraction charges can contribute positively to the first three of these fields and, below, it is shown how this can be done, using what we shall call *full cost incentive charging*.

The preparation of a tariff scheme should take as its starting point the regulatory controls put in place by the catchment authority in carrying out its responsibilities. In the development of these controls, there is an important role for social cost-effectiveness analysis and social cost-benefit analysis. These techniques have a place, alongside environmental impact assessment, in comparing alternative regulatory options. The design of environmental regulation should be an economic process as well as an ecological one.

The tariff regime introduced to any specific catchment (or region) will be contingent on its physical geography, its habitats and species, its human settlement patterns, its economy and the power relationships which hold between various social and economic groups. So what we require are criteria for tariff design which can be applied in the appropriate way to any single area with all its unique characteristics. We propose three such criteria for full cost incentive charging.

First, the annual income from abstraction charges should be hypothecated to the environmental regulator such that, when added to the income received from discharge fees, fishing licences, navigation permits, etc., abstraction fee income is sufficient to cover all the state's capital and current account expenditures on environmental regulation, research and database development, compensation payments, etc. In providing the regulator with hypothecated income, it is likely to increase their relative power as an agency of government. As Kraemer writes of Germany since 1988: "On the whole, the water resource taxes contributed to capacity building within the water management administration in the German Länder and thus partly overcame the implementation deficit in water resource management". It will also provide a budget to finance the legal costs of modifying or terminating abstraction licences that threaten the hydrocycle or undermine nature conservation. Higher abstraction charges will also give a price signal to water companies to reduce their storage and distribution leakage between the points of abstraction and the user's gate.

Second, where the full cost abstraction tariff still leaves an excess of demand for abstracted water over its licensed supply, the charge should be raised so that market clearing takes place.

Third, specification of the components of the abstraction charge should provide incentives for abstraction behaviour that is economically efficient and that avoids environmental degradation. Price per unit volume should be invariant with total volume abstracted, unless there are countervailing economic or environmental arguments. Price should be discounted where abstractors recycle their off-take to surface or groundwater sources. (Discharge fees should be used to handle the water quality aspects of recycled water.) Charges should be higher for abstracted water of higher quality. Seasonal variations in effective rainfall and economic demand should be dealt with through the licensed volume provisions laid down by the regulator and by the market-clearing criterion. Charges should be higher for upstream sources and for abstraction in locations where species and habitats are more threatened by abstraction.

The institutional framework that would be most appropriate for full cost incentive charging deserves discussion and my ideas here have been strongly influenced by Karin Kemper's *The Cost of Free Water*. The basic approach is a negotiation model in which there exists a public-sector catchment agency that, through a negotiating forum, develops its policies with the advice of water companies, direct abstractors, environmental organizations and water user associations representing the domestic sector, irrigated agriculture, mining and manufacturing, etc. The original prototype for such negotiation models is the French water parliaments. Central government retains the statutory right to determine which public and private institutions may enjoy the right to abstract water, on what scale, in what locations, at what time of year, at what price-but it delegates such rights to the catchment agency. The agency, with the assistance of its partners in the negotiating forum and with a full understanding of existing customary rights, then assigns formal abstraction rights on a time-limited basis to specific abstractors or groups of abstractors. The time limit would be 10 years, let us say, rolled over each year except where the agency allows the licence to expire. Where these 10-year rights need to be modified or rescinded for hydrological, environmental or economic reasons, compensation would be payable to the abstractor so affected. Rules would also exist to address third party impacts. Abstraction rights assigned to abstractors could be freely traded provided the agency had approved such transfers in the light of their social, economic and environmental impacts. Full cost incentive charging would be the basis of the price paid by abstractors for their water. It has the objectives of:

- Underpinning environmental regulation with a hypothecated income source.
- Requiring abstractors (and therefore final users) to cover the full costs of regulation.
- Bringing the quantity of water demanded by abstractors into line with regulatory limits.
- Giving price signals that promote both allocative efficiency as well as abstraction practices that avoid damage to riverine eco-systems.

No charge would be made for abstractions below a minimum scale–the transaction costs of such charges would be high compared with the volumes abstracted.

Arrangements would be made to monitor abstraction with respect to its location, time and quantity, as well as to invoice abstractors, to collect the charges owed and to enforce all agreements. Such transaction costs would be included in the full costing tariff of abstraction charges. The creation of this institutional framework imposes social, economic and political costs on the parties concerned, structural costs of change both real and financial in their nature. Therefore, the negotiating forum may agree that it is sensible for full cost incentive charging to be phased in gradually.

Irrigation Management Transfer on Mexico

Local associations of water users serve the same functions as irrigation agencies, but on a very localized level. In some countries such as Nepal and the Philippines, the major portion of the irrigated area is managed locally, through village-based water user associations. Typically such associations manage very small irrigation canals that were constructed by the users, perhaps centuries ago and the associations have grown up around the need for operating and maintaining these canals. In many developing countries such traditional canal systems have been the target of modernization efforts by government agencies, often funded through international development assistance. The relatively crude physical infrastructure of many of these canal systems was rebuilt and absorbed into the domain of the government agency, which replaced the management functions of the indigenous water user associations.

Today, local water user associations are recognised as resources of "social capital" which will have an increasingly important role to play in the coming decades. Instead of absorbing such associations into the state organization, the capacity of associations is developed so that they can improve the performance of their own irrigation and drainage systems. This participatory approach was re-invented in the Philippines in the 1970s and has since evolved into a global trend that combines participation with various degrees of privatization. In many developing countries a major focus of this process is the institutional challenge of establishing new water user associations which can serve the management functions previously handled by government.

A particularly fruitful source for a case study of this trend is Mexico. With 5 million irrigated hectares it was the seventh largest irrigator in the world in 1995. It began the irrigation management transfer process as early as 1989 and the experience has been extensively researched. Finally, Mexico has for a number of years been considered as a paradigm for this form of institutional transformation. The case study draws heavily on the work of Kloezen, Garcés-Restrepo and Johnson.

The broad political and economic context within which Mexico introduced irrigation management transfer was central government's neoliberal response to the economic crisis of the 1980s and the unsatisfactory performance of the irrigation districts under the management of the National Water Commission. Transfer has been a top-down process motivated by the international development banks, the main objective being to reduce public expenditure on irrigation OM&M whilst promoting greater user participation in irrigation management. It has been accompanied by a new National Water Act, revision of the Constitution to give legal foundation for the privatization of the *ejidos* (land reform communities) in all irrigation districts and termination of guaranteed crop prices and subsidized credit. "Dismantling the public sector, including the public irrigation sector, would not have been possible without the commitment at the highest political levels to reduce staff working in the public sector."

The institutional transformation has been rapid and radical. In its first phase the National Water Commission retained management of the headworks and the main canals. Water user associations took over financial and managerial responsibility for OM&M below the main canal. In the second phase the Commission is left with management of the reservoirs and surface-water-pumping stations; a District Federation of the user associations takes on the OM&M of the main canals. These associations are non-governmental local entities for collective action.

The high speed of the process and the low resistance of farmers have surprised many observers. It seems that transfer was preceded by other neoliberal reforms in agriculture. Farmers knew the Commission's traditional OM&M was to be terminated and OM&M had in any case declined in quality. Reform came first in the larger districts in the north, where many large private producers were known to support change. Lastly, the programme built on a strong organizational base-the *ejidos* themselves and the private farmers' cooperatives and unions.

Government informed farmers of the impending change–there seems to have been no serious consultation–and instructed them to select their delegates to the new user associations. Thereafter, delegates to each association elected from among themselves a president, a treasurer and a secretary. The Commission worked with the new user groups for 6 months or more from the time of transfer and provided their leaders and technical staff with extensive training in OM&M and financial management. The Commission also granted concessions to the associations to use its machinery and equipment so that there would be no capital expenditure required in advance.

Prior to transfer the National Water Commission was wholly responsible at the district level for the planning of annual and seasonal water allocations. Similarly the Commission employed the heads of the irrigation units into which each district was divided. These units were more or less independent hydraulic blocks with sizes ranging from 3,000 to 20,000 ha. Unit heads and their channel-keepers were responsible for daily OM&M at all system levels down to the farm inlets. Farmers paid their fees at the unit office and were given water by the channel-keeper.

After transfer, hydraulic committees were introduced at district level for allocation planning; these committees were composed of representatives from the Commission, the state government and each user association in that district. At the same time the district "units" were replaced by "modules"-two or more per unit–and it is at this level that the new water user associations manage. They collect irrigation water fees directly from farmers and employ their own module managers, channel-keepers, maintenance personnel and administrative staff. Based on the volume of water a module takes from the Commission and on the proportional amount of main infrastructure serving a module, each association must pay a percentage of the total fees it collects to the Commission. The specific subject of this case study is the Alto Río Lerma Irrigation District. Located in the State of Guanajuato, it has an area of 113,000 ha. There are some 24,000 farmers of whom 55 per cent are from the *ejidos* and 45 per cent are private growers.

The average landholding is 5 ha. Private holdings average twice the size of *ejido* holdings. The district's climate is sub-humid with annual precipitation of 750 mm and potential evapotranspiration of 1,900 mm. Eighty millimetres of rain falls in the winter season from November to April and 670 mm falls from May to November. Average temperature is 19°C and relative humidity is 60 per cent.

The Alto Río Lerma Irrigation District enjoys access to both surface water and groundwater. The river has four storage dams with a combined capacity of 2,140 mcm, as well as five diversion dams. The distribution network has 475 km of main canals and 1,660 km of secondary and tertiary canals, as well as 1,030 km of drainage canals. There are 1,710 deep wells exploiting three different aquifers with a total annual recharge of 500 mcm. Seventy per cent of the irrigated hectarage uses surface water and 30 per cent uses groundwater. Wheat and barley are the main crops in the dry winter months. Sorghum, maize and beans predominate in the wet summer season. All farmers grow vegetables and the private growers do so for the export market.

At the start of each agricultural year in November the hydraulic committee decides on the total area that can be safely irrigated in the district and by each of the eleven modules. This total area is derived from the combined volume of water in the four storage dams. The committee next sets the number of times that irrigation services can be delivered to each farmer in each season, typically 3-5 times in the winter and once in the summer. Each module receives its allocated share of surface water over the year and, given the module's ability to restrict the area irrigated by users, farmers can request water at any time within the constraints of the seasonal maximum. Farmers pay a fee to their water user association prior to each individual irrigation and get a receipt that has to be shown to the channel-keeper before water is allocated to their fields in terms of the charge categories.

Kloezen and his colleagues show a keen awareness of the political and macroeconomic context of irrigation management transfer as well as a familiarity, for example, with Fox's (1996) sociological work on building social capital from below. However, the language of common property resources, freeriding and transaction costs is absent from their reports.

What we do learn is that the number of staff engaged in governance, OM&M, administration, monitoring and evaluation *rose* by 13 per cent between the "before" and "after" transfer periods of 1992 and 1996. The proportion of persons employed by the National Water Commission dropped from 100 per cent to 38 per cent. The associations feel that the Commission staff numbers are excessive and that:

> An unspecified percentage of these...staff are residual personnel that for political and labour-union-related reasons remain within the agency with no specific task. This has been one of the major reasons why the modules wanted to create the [District Federation of the User Associations] which will take over management of the main system. They feel that this will be much more cost-effective than the current arrangement.

As already indicated, the water user associations hire all their own staff. In doing so they have shown reluctance to employ ex-Commission personnel. The associations argue that Commission channel-keepers and other staff were often out of control, poor workers, unaccountable, given to rent-seeking behaviour and were likely to involve the trade unions in module management in a divisive manner.

To what degree do farmers accept that the shared use of the resource is now equitable? The importance of the new, district-wide, hydraulic committees should be noted, but the allocation planning method has not changed at this level. However, there is evidence that correspondence between the volume of water assigned to modules and the actual volumes received by them is better than before transfer. This is probably because every association was represented on the hydraulic committee, with each module insisting on receiving the volume it had been assigned and had paid for.

With respect to individual farmers, a detailed, post-transfer study shows that in the sample area "there is no clear bias towards head- or tail-end farmers and that all farmers receive sufficient water to meet crop requirements". In a sample of farmers, 34 per cent said water distribution amongst them was poor before transfer and good after it, compared with only 15 per cent who held the reverse opinion.

Probably the most important contribution to equitable distribution of the resource has come from the associations' control over channel-keepers. Bribes exacted by Commission employees under the previous regime were impossible for farmers to eradicate. The downward shift in power to water users now gives them the control over rent-seeking behaviour that they lacked.

To what degree has infrastructural operation and maintenance been improved? With respect to operation, the post-transfer study indicated an improvement in water adequacy at field level, in the timeliness of water delivery and in access to the channel-keepers. In the sample of farmers, 40 per cent said the service provided by the channel-keepers was poor before transfer and good after it, compared with only 14 per cent who held the reverse opinion. Eighty-three per cent of farmers believe that the water user associations should now retain responsibility for operation of the main and secondary systems.

The transfer process included free concessionary use of Commission machinery such as draglines and hydraulic excavators. This was a flying start for users. Significantly, some of this equipment was in serious disrepair so the modules bought twenty-nine pieces of new heavy machinery from their user fee income and also received equipment from a World Bank programme. Maintenance expenditure in constant peso prices almost doubled in the before- and after-transfer comparison. There was an approximately threefold increase in the desilting of secondary canals and drains. Farmers' perceptions of the condition of the irrigation and drainage network were that the network had improved considerably.

To what degree have negative externalities arising from individual farmers' actions declined since the user associations were set up? Kloezen and his colleagues address this question only with respect to over-abstraction of groundwater. By customary practice, modules exclude from the surface-water supply those areas that have access to wells. But groundwater users are permitted to use the canal network to distribute their pumped supplies. Unfortunately the aquifers in the district are definitely being over-exploited by about 20 per cent of annual recharge. By 1996 the water table was falling by 2-5 m annually. Groundwater mining does not appear to have become worse since transfer, but transfer has been unable to diminish it.

To conclude this case study, the new arrangements for the financing of capital and current costs are examined. We have already seen that the association pays fees to the Commission for its services, that the association charges the farmer for each irrigation, that there has been a decline in bribes paid to channel-keepers and that charges to farmers have been used not only for prime costs but also for equipment purchase.

Forty per cent of farmers surveyed report that payment procedures are now less cumbersome compared with only 2 per cent holding the reverse opinion. Sixty-nine per cent stated that bribery of channel-keepers had been reduced. But many farmers still come to the module office to complain of keepers demanding private payments and this has often led to workers being dismissed. Keepers are also rotated between module sections to prevent the growth of patron-client relations. Nevertheless as many as 30 per cent of farmers surveyed were willing to admit that they bribe keepers to irrigate more than the farmer's entitled area or to get water at different times from those programmed. So the indirect charges are higher than the official ones-as are the wages of keepers.

Under the Commission, charges were way below prime costs, let alone the full cost of the service. To prepare the way for a more self-reliant organization, the Commission increased the charge by some 400 per cent 2 years prior to transfer. It then remained unchanged in nominal pesos through and after transfer. But inflation halved the real value of the charge in the first 2 years of the associations' existence.

What is undoubtedly impressive is that in terms of the ratio of fees actually collected to actual OM&M expenditure, both given in pesos at current prices, the rate jumped from an average of 50 per cent in 1989-91 to 123 per cent in 1993-96. These facts remind us that in irrigation economics the estimation of the cost of water to farming families and agribusinesses should be based on the payments *actually* made, not on those that are *supposed* to be made. Kloezen and his colleagues never attempt to measure what proportion of the full cost of irrigation services is met by farmers' actual payments, but it seems likely to be considerably less than 100 per cent. This raises the question: "What organization in future will be responsible for infrastructural investment and rehabilitation, how will it raise the required finance and how will these costs be repaid?"

For the 1996-97 winter season the associations used the hydraulic committee to raise the irrigation fee by more than one-third. They also pushed the Commission to transfer OM&M responsibility for the main canal to a newly created District Federation of the user associations. In 1997 the average percentage of farmers' irrigation fees paid to the Commission was correspondingly slashed from 25 per cent to 9.5 per cent.

Tradable Abstraction Rights in Australia

Up to this point, consideration of abstraction has covered the right to abstract and the charges that a catchment agency might levy on farmers for the enjoyment of that right. Tradeable abstraction rights (TARs) have been discussed only briefly-although long enough to recommend that, with appropriate safeguards, such rights should be widely introduced. This case study reviews the introduction since the early 1980s of TARs within irrigated agriculture by the State of Victoria in Australia. The case study updates by 4 years one that we published in 1997.

In every catchment, some users of water will value their abstraction rights more than other users, for example, because they employ water more productively. In that situation, trading in such rights can be financially attractive to both the existing owner of such rights and the party that wishes to see them transferred. To understand trading in abstraction rights we need to come to grips with the relevant law, with differential valuations of water and with the legal, hydrological and engineering means for implementing transfers. In this field, valuable work has been carried out by Robert Hearne and William Easter, in Chile, although when they were writing their report less than 1 per cent of all abstractions operated through such water markets there.

In Australia, the practice is now extensively developed and there the main objective of TARs has been the more efficient use of scarce irrigation water. The Australian water economy is said to be in its mature phase, where the long-term average total cost function is rising sharply, where there is intense competition for existing supplies and where the hydrosocial infrastructure requires costly rehabilitation.

Victoria is one of the constituent states of federal Australia and lies in its south-east corner. The dominant physiographic feature is the Great Dividing Range, running east-west in the eastern two-thirds of the state. This creates a rainshadow in northern Victoria, where average annual rainfall is 450 mm in the Goulburn-Murray Irrigation District (GMID). This is the largest irrigation area in Australia, covering some 820,000 ha and it represents the bulk of water trading in the state. Irrigation is sourced from reservoirs on regulated surfacewater flows. Meat and dairy products are the principal agricultural output; higher-valued crops such as horticultural products are constrained by the red-brown soil type. Surface-water resources have been extensively developed in the past and any further supply growth would require inter-basin transfers. The GMID is principally designed to provide security of supply during a prolonged series of drought years. The main water dams are operated on a carry-over basis, where water is accumulated in years of high river flows for use in drier years.

Australia's modern history of property rights in water begins in the 1880s, when Victoria introduced new water laws based on the recommendations of a Royal Commission chaired by Alfred Deakin. He proposed that water allocations should be tied to the land, that rights to water should be vested in the British Crown and that allocations to landholders should be the responsibility of the state governments. Riparian owners retained limited common-law rights for domestic use, stock watering, gardens and a maximum of 2 ha of irrigated land for fodder crops. Over the next 15 years, similar legislation was introduced in the other federal states. Using the new legislation, the states ventured into large irrigation projects, providing water to farmers far below average total cost. Areas were established in which government constructed massive irrigation and drainage infrastructures, such as the Goulburn-Murray Irrigation District itself, the Murrumbidgee Irrigation Area in New South Wales and Riverland in South Australia.

The subsidy of water led to overuse and this was consolidated by the historical overallocation of water to irrigated farms. As argued in the work of Pigram and co-workers: This has its roots in the social objectives of past governments for the development and use of water resources. The overriding objectives were an equitable distribution of water among farms and the promotion of regional development and closer settlement of inland areas. All farmers were considered to have an equal right to the available water, irrespective of how much was needed to irrigate the proposed crops to be grown on the farms, or of any consideration of how efficiently the water would be used. As water rights became capitalized into land values, individuals had the economic incentive to retain their entire water rights through demonstrating a history of use. In many cases, this actually translated into a history of over-use.

By 1990, the arrangements in the GMID were that irrigators paid for a water right based on both the amount of land they held that was suitable for gravity-fed irrigation and the irrigation district in which they were located. An annual charge applied to the water right, invariant with the volume of use.

Subject to availability, "sales water" was also available, at a volumetric charge based on the water right charge. These water rights in their specific form were assigned by a licensing system for a fixed period of 15 years, with an expectation of reissue. The authorities could vary the volume licensed, usually in times of shortage. However, before 1987-88, no arrangements existed for *transferring* farmers' abstraction rights.

An Australian interest in TARs (also known as tradeable water rights or transferable water entitlements) began to surface in the mid-1970s. For some, the introduction of TARs seemed to offer a new flexibility to existing arrangements, with clear economic and environmental advantages. *First,* we have already seen that the maturing of the Australian water economy in recent decades is associated with high average total cost for the long-term supply curve. So, TARs offered a new direction that, by reallocating water supply, would reduce the pressure for aggregate supply expansion. This reallocative effect would not be merely within the farming sector. There was also the prospect of reducing agricultural overuse to open up supplies for a range of urban and industrial uses.

Second, supply reallocation within agriculture was a major objective. There was a widespread belief by the mid-1980s that TARs would switch water from lower to higher water-productivity uses in the farming sector. In Australia, at that time, the agricultural sector accounted for 80 per cent of total water use. In Victoria, each year, up to one-third of irrigators were using less than their full water right allotment. Specific switches into river red gum watering, salinity dilution and dairy farming were forecast. *Third,* environmental benefits were expected from transferability. The new policy promised to reduce the scale of infrastructure construction for inter-basin transfers, with all their negative externalities. Development proposals were meeting increased resistance from environmental groups. Moreover, reductions in agricultural overuse promised to have a positive environmental effect through a reduction in waterlogging, salinization and biocide dispersion. The redistribution of abstraction rights could have been sought by the administrative processes of the licensing system. But this would have been met by political resistance from the farms on which volumes were to be reduced and land values consequentially cut. Tradeable abstraction rights offered these farms a pay-off.

For landholders wishing to move out of irrigation, but to remain in dryland agriculture, transferability allows water entitlements to be sold separately from the land. Previously, the options for such irrigators were either to cancel their water rights licence (or not renew it) and get nothing for the right, or sell the entire irrigation holding and buy a dryland property elsewhere. Transferability can permit a more flexible retirement plan for an irrigator, or facilitate a long-term change in enterprise or financial structure of the farm business.

The fundamental requirement of a workable and efficient market in abstraction rights is a clear specification of property rights in water such that:

- Rights in land and rights in water must be separable.
- The volume of water that an individual or institution has available for transfer must be clearly stated, as well as any special conditions on its use-such volumes are likely to be conditional on rainfall, or surface flows or groundwater stocks.
- The right to transfer such water at a privately negotiated price must exist.
- The period over which such a transfer is deemed to be effective, temporary or permanent, must be known.
- The power of government to restrict or terminate abstraction rights at a future date must be known.

In Victoria, TARs were cautiously introduced in 1987-88 with a *temporary* scheme, before permanent transfers were considered. In the GMID the transfer period was to be for 1 year, only between irrigators and only within the same supply system. There was no volumetric limitation but stock and domestic allocations had to be retained. A state agency assessed possible third-party effects and could refuse the transfer if these were significant. The arrangements were not to affect significantly delivery and drainage channel capacity or salinity. Finally, the price was determined between buyer and seller but the agency's fixed fee was A$70.

A new Water Act in 1989 permitted *permanent* transfers of abstraction rights between farmers, with effect from 1991-92. Transfers out of agriculture were still proscribed. However, despite these references to "permanence" in trading, the state assignation of water rights through the licensing system, described above, meant that a purchaser of abstraction rights could not be assured in law that such water rights would continue indefinitely. The state government retained long-term flexibility in its management powers.

TARs impose certain transaction costs. In the GMID case they seem to be modest because of the legitimacy of the institutions, the support from farmers for the general approach to the new rights and the competence and honesty of Victoria's state administration. Incremental supply costs also exist if distribution and drainage structures cannot handle the increased water volumes. However, in Australia, water agencies have taken the easy administrative option of simply refusing transfers where existing capacities would be exceeded.

The outcomes of Victorian TARs in the early years after they were first introduced are considered next. The scale of change can be measured by the volume of abstraction transfers as a percentage of all abstractions. With respect to benefits, we are looking for: the avoidance of new infrastructure expenditure as a result of the redistribution of abstractions within and between user categories; an increase in agricultural productivity as a result of abstraction redistribution within farming; and a fall in waterlogging, salinity and pollution from run-off because of the reduction in agricultural overuse.

In the 2 years 1987-88 and 1988-89 the trading of abstraction rights in Victoria (most of it in the GMID) averaged 25 mcm, less than 2 per cent of total abstractions. Price per megalitre ranged from A$8 to A$20. Gross margins increased. But at this point the judgement was that "Despite widespread endorsement of the concept in Victoria, transfer activity has been sporadic in that state."

In Victoria by 1990-91 trade was still "insignificant compared to the total volume". In a mail survey of all 299 permanent water transfers within the GMID up to 1994, with a 63 per cent response rate from buyers, purchases made up between 0 per cent and 2.3 per cent of total allocations when the responses were grouped by district. What the survey showed was that sellers were releasing sleeper water (i.e., unused allocations) and that farmers in financial distress were also selling. In a smaller number of cases, farmers either wished to cut their irrigation agriculture, or to farm without irrigation, or to retire. In respect of purchasers, the single most important reason for buying water, applicable to 65 per cent of respondents, was that they "wanted to secure existing crops against future drought". Stringer suggested that in Victoria as a whole in 1991-94, the average percentage of allocated water traded in each of those years was 0.33 per cent. In the Murray region as a whole in 1993-94, the volume of transfers was 17 mcm for temporary transfers and 8 mcm for permanent transfers. For 1994-95, these figures changed dramatically. Temporary transfers soared to 265 mcm whereas permanent transfers fell to 2 mcm. At this point it could be argued that by the mid-1990s TARs in Victoria had created a space for single-season switching of abstraction, most active in years of low rainfall.

By the year 2000 Australian statistical reporting of TARs had become much fuller and more informative as a result of work by the Australian National Commission for Irrigation and Drainage (ANCID 2000). Its *Benchmarking Report* contains data for nine Goulburn-Murray systems known as Murray Valley, Shepparton, Central Goulburn, Rochester, Pyramid-Boort, Torrumbarry, Nyah, Tresco and Woorinen. In all of these except the last the percentage of surface-water irrigation supply points that are metered lies in the range 87-98 per cent. A broking service exists in all nine systems to facilitate the trading process between users. In 1998-99 temporary trading took place in all nine systems, averaging 11 per cent of total abstraction entitlements for that year. Six of the nine systems reported permanent trading that averaged 2 per cent of entitlements in those six areas.

From the perspective of this book, the material on why one set of irrigators sell their TARs and others purchase them is of particular interest. A substantial paper by Björnlund and McKay (2000) throws some light here. The article is based on more than 300 telephone interviews conducted in 1994-96 with buyers and sellers of permanent TARs in the GMID.

The authors state that the main use of irrigation water is in the cultivation of pastures for dairy cattle. Some activity is found too in horticulture and viticulture. Mixed grazing and cropping is said to have low-value water use.

Purchasers of tradeable abstraction rights were shown to be seeking to ease their reliance on annual sales water and temporary water market purchases. TAR sellers are smaller than buyers in terms of irrigated area; water trading can be seen to be integrated with a long-term consolidation and amalgamation of farm properties.

Sellers were asked how they used the cash proceeds of their permanent sales-note that a permanent sale does not imply the sale of a farm's entire water entitlement: 10 per cent was invested in laser grading, 5 per cent in improved drainage and another 5 per cent in other irrigation equipment. Fifty per cent of the sample said the money was used for general consumption purposes, related or not to the farm as an enterprise;

21 per cent went to debt reduction; 6 per cent to non-irrigation equipment; and 3 per cent for the purchase of another farm.

Clearly the subject is ideally suited to an irrigation economics case study, for both the permanent and temporary markets, into the differential motivation of buyer-seller *pairs*.

A BLOCKED CHANNEL

One aspect of this blocked channel concerns payments for abstraction rights. We have given a very full account of the criteria that can be used by a state authority in constructing an abstraction tariff for farmers and the relevance of such a tariff to effective catchment management. But there is no accompanying case study to test these ideas–merely brief, illustrative examples. In contrast there is an extended discussion of the recent history of tradeable abstraction rights in Victoria, but the theoretical material is adhoc and fragmented.

Despite this the research reports of Wim Kloezen, Carlos Garcés Restrepo and Sam Johnson seem to be wholly successful in coming to grips with the dynamic institutional changes of this catchment. Perhaps it is because they were a team with varied disciplinary backgrounds, which is so important in water resource planning. Perhaps it is that irrigation management transfer is quintessentially a compound of private economic interests and political power set within a given hydrological context–and that Kloezen and his colleagues have a good grasp of hydropolitics. In contrast, the language of many economists is so distant from the complex drives of individuals, families and organizations and so wedded to a vision of the determining interplay of measurable quantities, that they fail to understand the nature and exercise of power.

Chapter 8

Drainage Service: Demand and Supply

THE DEMAND FOR DRAINAGE SERVICES

The economics of land drainage in agriculture–a subject that is the poor and neglected cousin of the economics of irrigation. Land drainage can be defined as those processes, natural and social, by which flows of water on the land, at both surface and sub-surface levels, are conducted to other locations, such as lake, river and aquifer.

The outcomes of drainage activity are not necessarily confined within the boundary of a single farmholding and therefore, like wastewater collection and disposal, drainage services may be a *public good*. A public good can be defined as any good where the benefit derived from it by one consumer does not diminish the benefit derived by consumers in general. The standard textbook example is a country's armed forces. Since no consumer privately appropriates a public good, it is often difficult and sometimes impossible to price or sell it in a free market.

In the absence of human society, the drainage of rainwater is a purely natural phenomenon. But with respect to the drainage of irrigation water and rainfall in farming areas we are dealing with flows that may require human intervention, so that we can legitimately speak of the demand for and supply of drainage services. The same holds true for urban drainage.

The water inputs that source drainage water in farming are rainfall itself as well as the irrigation system's storage points, distribution network and irrigated fields. Drainage water's constituent flows are leaching applications, escapes, seepage, deep percolation and surface run-off. Drainage water's outputs are non-beneficial evaporation and transpiration, returns to the catchment's hydrological resource and the pumped flows that make up the fifth category of water appropriated for irrigation.

The demand for drainage services by farming families and agribusinesses is found whenever gravity-driven 'natural drainage', without any human operational or infrastructural intervention, does not meet farmers' needs effectively in comparison with drainage services that have been engineered. Farmers' need for drainage services takes a number of forms. *First,* rainfall and irrigation water flows can collect on the land's surface at various points in the farming area. This may be due, for example, to unevenness in field level or to soil types such as clay that inhibit surface water from percolating downwards. As a result, the ground becomes waterlogged and the excess of soil water at the plants' rootzone constrains their growth and reduces tonnage efficiency. Waterlogging also occurs where the water table is shallow refer to water tables as close as 50 cm to the land surface in Haryana, India, where seepage of hundreds of millimetres per year from distributaries has caused extensive waterlogging.

Second, drainage water can be reused for irrigation if that water is channelled into ditches or pipes and pumped up to field level. The additional irrigation flow is used to increase crop turnover. Here we have the fifth category of the irrigation cycle's base supply.

Third, drainage water is used for artificial recharge of the local aquifer. The additional stored volumes maintain crop output, by means of groundwater pumping, whenever rainfall is insufficient.

Fourth, irrigation water invariably contains dissolved salts. This may already be the case at the point of appropriation of the base flow, it may be introduced by farmers' application of agrochemicals, or salts may be dissolved as the surface flow of irrigation water is distributed to the plants' rootzone. The evaporation of irrigation water at the field level can increase salinity up to and beyond the concentration where it begins to inhibit plant growth. Once again, tonnage efficiency falls. Salinization also occurs where the water table is shallow, 2 m or less below the soil surface. The upward capillary movement of groundwater and its evaporation leads to salt accumulation in the soil profile. In the cases of both irrigation water and groundwater evaporation, the processes of natural drainage lower tonnage efficiency. Here planned drainage can be beneficial, including the case where water is supplied to fields and then drained precisely in order to leach salts from the soil profile Farmers are likely to practise leaching at times of the year when the opportunity cost to them of the water thus used is lowest.

To sum up, the demand for drainage services in agriculture is fourfold: to prevent waterlogging, to source one of the irrigation cycle's five base flows, to facilitate aquifer recharge and to combat soil salinization. The first and fourth of these raise tonnage efficiency whereas the second and third augment the availability of irrigation water. In all four cases, turnover per holding is raised, provided the increase in quantity supplied does not face inelasticity in demand.

Van Hofwegen and Svendsen report that about one-third of global cropland–500 million ha–is not adequately drained by natural processes.

In contrast, in Asia, Europe and North America engineered drainage has contributed to raising food production on about 150 million ha. But such drainage can destroy natural wetlands and, as a result of the growth in power of the environmental conservation movements, new reclamation has slowed considerably. But much existing agricultural land still suffers from inadequate drainage in both rain-fed and irrigated areas-South-east Asia in the monsoon seasons is an example. "In arid and semi-arid areas some 20 to 30 million hectares of land is suffering from water logging and salinity due to irrigation or high canal percolation rates."

It is worth noting the important distinction between surface and sub-surface drainage. Their technology, infrastructural costs and effects can be strikingly different. In Finland, for example, without proper sub-surface drainage of heavy clay soil, waterlogging due to low hydraulic conductivity of the surface soil and especially the subsoil leads to abundant surface run-off. In turn this induces soil erosion and phosphorus losses.

PAYING FOR DRAINAGE SERVICES

The demand for irrigation services and the demand for drainage services are agronomically distinct, one from the other, even when those services are interdependent in various ways. The single feature that most distinguishes drainage from irrigation is that a drainage service is necessary to cope with the surface and sub-surface flows of rainwater even in the complete absence of irrigation. Remember that at the global scale about two-thirds of all agricultural land is rainfed and enjoys no irrigation supply.

However, with respect to drainage service cost to the farmer, there are marked similarities with irrigation. Drainage services may be delivered at no cost to the cultivator or that cost may be borne directly by farmers themselves. There may be a volumetric fee charged to the farming family or agribusiness such as is possible in the case of pumped vertical drainage. In irrigated areas it is commonly the case that the supply of irrigation services and the supply of drainage services are carried out by the same institution. Where this happens, both are likely to be financed by a single payment mechanism.

Just as with irrigation, we can speculate on the elasticity of demand by farmers for drainage services. *P* would be the price of each cubic metre drained and *Q* would be the volume of water drained. But drainage services are likely to be seen by farmers more as an agronomic necessity rather than as an economic commodity. Nevertheless, Carruthers and Clark cite a counter-example. They state that most cultivators in Pakistan were share-tenants, "who are, moreover, moved around by the landlords to different plots each year". Share-tenants have a short time-horizon with respect to investment in farming; in this specific case they showed little interest in field levelling, an important technique for improving drainage and leaching.

THE SUPPLY OF DRAINAGE SERVICES

At the heart of drainage engineering we find a means whereby the natural, gravitydriven flows of rainfall and surplus irrigation water across and beneath farmland are substituted by man-made channels powered by gravity or pumps. As a result, from a civil engineering and hydraulics perspective, the supply infrastructure of drainage services bears a striking similarity to the abstraction and distributional infrastructures of irrigation supply.

The channels are dug ditches without lining or lined channels or pipes. The efficiency of axial flow pumps in lifting large volumes of water at low pressure through a few metres, a common case in land drainage. Deep tube wells may be employed, as in Pakistan's Punjab province, to pump water for vertical (rather than horizontal) drainage in order to reuse the flows, perhaps after adding river water to reduce the salt content of the conjunctive supply. Vertical drainage may also take place to lower the water table and drive salts in the soil below the rootzone of the plants, where they can be harmlessly stored.

The common features of the civil engineering of land drainage and that of irrigation water abstraction and distribution have important consequences in economic terms. Land drainage networks may require major capital expenditure and capital financing. In the short term there are substantial economies of capacity utilization. In the long term, from the planning approach, we can also expect economies of scale because of the indivisibility of drainage infrastructure provision. Lumpiness is also likely to be a feature of drainage networks. The double punch of indivisibility and lumpiness means that mediumterm and long-term planning for drainage offers considerable benefits.

Drains require continuous work to remove debris of all kinds, to dredge silt and to cut back plant growth in and on the edge of watercourses. The mowing of bankside vegetation provides control of weed growth and encourages a healthy root system, stabilizing the banks and giving some protection against erosion and slippage. Aquatic vegetation assists oxygenation of drainage water and channel stabilization and also provides habitat for fish and invertebrates. But in excess amounts such vegetation holds water back, raises the water level, encourages siltation and leads to flooding. In Norfolk channels are desilted only when 300 mm of silt has accumulated. A channel's centre section is desilted over its entire length but the two margins are left untouched. "This method is preferable from the land drainage standpoint as it allows continuous unimpeded central flow and will also assist future conservation by the creation of pools and shallows at the water's edge". Hydraulic excavators, draglines and tractor-mounted mowers may be used to automate the maintenance process. Sluice gates require sluice-keepers or may be operated remotely using telemetry systems for monitoring purposes.

Currently a strong interest is shown in controlled drainage strategies. These are practices that allow the farmer to trigger drainage only after the groundwater level in a field has risen so far as to threaten crop yields or prevent salt leaching. The associated reduction in drainage flow under such a regime permits a lower rate of abstraction for irrigation purposes. To the degree that unnecessary overdrainage is reduced, more water becomes available for downstream irrigators or for non-agricultural abstractors. The two main methods for drainage control are:

- The blocking/unblocking of lateral or collector pipes using a purpose-designed device. Here the water table rises and falls in response to irrigation releases and switching drains off or on. Farmers in some areas have used mud and straw as the control device but this can lead to system blockages.
- The placing of a (fixed or adjustable) weir in the drainage ditch or sub-surface drainage system. Here the drainage flow takes place only when rainfall or irrigation inputs bring the water level above the top of the weir.

It is suggested that controlled drainage can be beneficial in many arid and semi-arid regions of the world where water tables are high, such as Egypt, India and Pakistan. In the Nile Delta controlled drainage trials under rice have achieved large water savings and farmers are being encouraged to adopt the technique. In 2000 Catherine Abbott and her colleagues at HR (Hydraulics Research) Wallingford and the Drainage Research Institute, Cairo, were carrying out field trials on controlled drainage of dry-foot crops to assess potential water-saving benefits. The work will culminate with the development of predictive tools to help assist in evaluating the costs and benefits of this approach.

Drainage Water Quality

Modern agriculture makes extensive use of biocides and fertilizers. The former protect cultivated plants from a great variety of natural pests, the latter promote plant growth, particularly by means of increasing the supply of the macronutrients nitrogen, phosphorus and potassium at the rootzone. As a result, the farmland drainage water that returns to the hydrological resource is more or less polluted.

The introduction of synthetic chemicals began in the 1940s–at least in the UK. Since that time many, many hundreds of products have been developed for use, particularly in the case of fungicides, herbicides and insecticides. Their use varies considerably from region to region and in the more developed farming areas such as the UK it is not uncommon for six or seven different chemicals to be applied in a single season.

In most countries our understanding of the scale and type of pollution is deficient because of the variety of chemicals and the costs of water quality monitoring. For the same reason, the impacts on human health and ecosystems is not well understood. This is exacerbated by the likely difference between short-term and long-term impacts.

As Falconer (1997:16) points out: "An important distinction must be made between point sources of contamination (e.g., from yard or storage spillages and careless or illegal disposal) and nonpoint sources, such as the diffuse emissions resulting from normal, legal field use ...chronic and diffuse problems are generally not reported due to their invisibility." The policy measures that have been adopted to mitigate or eliminate the agrochemical pollution problem are varied.

They include:

- The promotion of organic farming-Keveral Farm is an example.
- Bans on the manufacture and use of specified products.
- The production of low-dose and less persistent chemicals.
- The levy of pesticide taxes as well as payments for the non-use of agrochemicals.
- Information and training for farmers on the safe use, storage and disposal of products.
- Specification of exclusion zones for agrochemical use around aquifer recharge areas, along watercourses and on field margins;
- The setting of minimum water quality standards such as for drinking water.
- The introduction of more comprehensive and more effective freshwater treatment.

Drainage water is an important medium for the diffusion of agricultural pollutants beyond the farming sector and its particular concern for soil salinization. The policy issues concern either the agricultural use of agrochemicals or the standards and technologies for the treatment of abstracted water destined for households and industry.

Thus the subject area, however vital, essentially falls under the economics of agriculture and the economics of freshwater treatment, outside the field of the economics of irrigation and drainage. For that reason, in this book fertilizer and biocide contamination of water by farming families and agribusinesses will be addressed only in so far as it influences the efficiency of irrigation water use.

Farmland Drainage

The county of Norfolk is located in the east of England and the waters of the North Sea wash its coast. Norfolk is renowned for its low topography, the fertile soils of its farmland and the rich diversity of its wildlife habitats. Annual rainfall is plentiful so that, with the flatness of the terrain, the inhabitants of the county for thousands of years have had to drain the land in various ways. This is of special importance to farming families and agribusinesses in order to prevent waterlogging and flooding. Important crops are wheat, barley, rye, oilseed rape, linseed, potatoes, parsnips, sugar beet, other vegetables and salad crops.

The advantages of cooperation in providing drainage services have been recognized in the area since at least the time of the Roman occupation of Britain. The modern history of drainage dates from the 1930 Land Drainage Act that required internal drainage boards (IDBs) to be established in low-lying areas of England and Wales to provide a common service. In each designated drainage district a board was formed from members elected by local ratepayers and local authorities. In the case of Norfolk a step of major importance was taken in 1967 with the creation of the King's Lynn Consortium of Internal Drainage Boards ("the Consortium"). Initially this brought together eight IDBs.

An important development took place in 1991-94 with successive Land Drainage Acts imposing environmental and conservation duties upon IDBs and providing for codes of practice on these matters. By the year 2000 the Consortium acted for sixteen IDBs. Its area in 2000 was 160,000 ha, it had forty-three pumping stations and it was responsible for 3,500 km of watercourses. The IDBs' locations are closely related to the position of the 20-feet (6.1 m) contour in Norfolk. The core duties of an internal drainage board are to secure the efficient working and maintenance of existing drainage works in its district and the construction of new works as may be necessary.

There are powerful economic and political reasons for the IDBs to work together within a consortium. Short-term and long-term economies exist in infrastructural provision. For example, it becomes financially feasible to purchase machinery such as a £100,000 hydraulic excavator used for drain construction ("drain" is the generic term used by the Consortium for its drainage channels). The equipment is fully used and its size avoids the costs of indivisibility. Similarly short- and long-term economies are to be found on the OM&M side, as well as with planning, finance, information technology and civil engineering skills. A single drainage district, for example, could hardly contemplate employing on a full-time basis its own graphical information system (GIS) expert or its own environmental scientist. The Consortium's total employed staff in

2000 was fifty-three–made up of eight professionals, six administrative staff and thirty-nine employees in the civil works direct labour force. The presence of a group of full-time professionals in the Consortium's headquarters with an annual expenditure of about £5 million and responsibility for 160,000 ha also gives the group political clout with bodies such as the Environment Agency, the Ministry of Agriculture, Fisheries and Food (MAFF), the Department of the Environment, Transport and the Regions, the Broads Authority, English Nature and the Royal Society for the Protection of Birds.

The specific focus of this case study is the West of Ouse Internal Drainage Board, located in the fenland area of west Norfolk between the towns of Wisbech and King's Lynn. West of Ouse is as flat as a pancake, consists entirely of farmland and about twenty villages, measures some 13,370 ha and is made up of land largely reclaimed from the North Sea between AD50 and 1978. The Environment Agency's long-term average rainfall data for the period 1961-90 varied between 553 mm and 595 mm across the seven points where it is measured.

The modal value is 579 mm. Over the same period Meteorological Office data show that effective rainfall averaged approximately 149 mm. The scale of irrigation in the area is not great and so drainage is primarily of rainfall run-off in order to prevent waterlogging. The water table is close to the land surface in this area. There is no reuse of drainage water for irrigation purposes. In general in Norfolk, irrigation takes place most on the higher, silted lands.

The drainage infrastructure is 620 km in length, almost entirely composed of excavated, unlined earth drains. One-third of the area, in the north-east, drains by gravity to five outfalls located on the River Great Ouse. The other two-thirds of the area, in the south-west, drains by gravity to Islington Pumping Station whence the drainage water is pumped up to One-Mile Drain from which it flows by gravity into the Great Ouse, which is still tidal at this point. One-Mile Drain also acts as a storage reservoir prior to discharge at low tide. All the outfalls to the river have pointing doors and are powered simply by the relative movement of drainage water and the river's tides. Islington Pumping Station has three diesel engines driving axial flow pumps and, since 1993, two submersible electric motors driving two mixed-flow pumps.

The Consortium staff maintains for the West of Ouse IDB an income and expenditure account as well as a balance sheet. This includes outgoings in respect of both farmland and village drainage. Just four items make up 87 per cent of the total. *First,* the Environment Agency precept of £176,000 is the payment due from West of Ouse to the Agency for the services it provides to this IDE. The most important of these services are the Agency's defence of the Wash coastline and the control of flooding from the River Great Ouse, primarily by means of embankments. *Second,* we have £121,000 in loan charge payments on the finance raised for various infrastructural works such as the Smeeth Lode channel and pumping plant. The UK's Public Works Loan Board (PWLB) sources such loans. *Third,* OM&M costs amount to £238,000; the item "Use of Consortium plant and labour" is West of Ouse's largest single entry. *Fourth,* the miscellaneous costs of administration total £114,000, the bulk of which is incurred at the Consortium's headquarters in King's Lynn.

The Consortium's expenditure on capital goods is funded by MAFF grants, Environment Agency grants, PWLB loans, internal funds and leasing arrangements with machinery and equipment suppliers. Leasing rather than outright purchase may secure Agency subsidies not otherwise available.

The final issue to be discussed in this case study is the way in which the Consortium's current expenditure on farmland drainage is financed. I believe the principles could be of great interest internationally. To begin with we should note that, whereas the Consortium's activities embrace both urban and farmland drainage, outlays in villages and towns are met from charges paid by local councils. Farmers are charged only for farmland drainage. It is also the case that in any single year, such as 1999-2000, farmland drainage costs are pooled across all the sixteen IDB members of the Consortium. The current costs of each single IDE, such as West of Ouse, can be identified, but that individual board's outlays are *not* the basis of that board's farmers' payments.

Instead, *all* the Consortium's current expenditures on farm drainage are aggregated and this sum is then divided for payment between all the farmers located in the Consortium's area. But what is the principle on which this division of pooled costs between farmers is made? At this point we need an historical perspective. In Britain the taxation of real property to generate local revenue is a practice of considerable antiquity and the Poor Relief Act of 1601 is regarded as the foundation of the modern "rating" system, as it is called. In origin, rates–taxes–were assessed on movable and immovable property, both domestic and non-domestic. In fact, the assessment of movable properties was so difficult that they were left out of account long before their formal exclusion by the Poor Rate Exemption Act of 1840.

A tax on immovable property is particularly suitable to the functions of the local state since evasion is extremely difficult and there is no problem in attributing the yield to even the smallest units of local government, as in the case of internal drainage boards. The tax can be levied in the household sector, on industry and in agriculture. Only the last case is considered here, although the taxation principle was originally the same for all three sectors. Within the relevant area each agricultural holding is separately recorded on a valuation list. In England and Wales the Local Government Act of 1948 gave the responsibility for valuation to the Inland Revenue, where local valuation officers carry it out. These officers then calculate for each holding the annual sum for which the land and buildings on it could be rented on the open market. It is assumed that the costs of insurance, maintenance and repair of the property is paid for not by the owner but by the tenant. This estimated annual rent is called the property's *net rateable value* and is set at a defined number of pounds sterling.

Specifically with respect to the net rateable value (NRV) of agricultural holdings–both the land and the farm buildings–these are based on what the Agriculture Acts call a "full agricultural tenancy". They were last assessed in 1991 and were carried out on contract for the Inland Revenue, mostly by private valuers but also by some public-sector district valuers. There has been no general reassessment since that time but provisions exist in the 1991 Land Drainage Act for a farming family to seek revaluation because of a change in the circumstances of a holding, for example the removal of greenhouses. The cost of the valuation process, a transaction cost, should not be neglected by any government considering introducing such arrangements.

Rates due for farmland drainage are set as a defined number of pence. For every 1 pound sterling in the NRV of a property, this number of pence is payable. (Since there are 100 pence in the pound, rates in pence can be equally well thought of as a percentage of the farm's NRV. That was not true before decimalization took place in the UK in 1971, at which time there were 240 pence in the pound.) In a simple case where the rate is 5 pence and Farmer Giles has a small property with an NRV of £1,000, the rates he would pay are £50, i.e., 5 pence in the pound multiplied by 1,000 pounds. As befits a wealth tax, the more valuable the holding in rental terms, the higher the rates and the greater the contribution made to drainage costs. The payment of rates is made by the occupier of the farmholding, for in its developed form it is regarded in law as a tax on the benefit of occupation of land and buildings.

In 1999-2000 the rate charged for land drainage in the Consortium's area was 6.77 pence. Because the NRV of farmland and farm buildings at the beginning of that year for the West of Ouse was £8.9 million, the total sum raised by the rates was £603,000. The specific costs on current account of farmland drainage in that year were £742,000. So the income from rates covered 81 per cent of drainage costs. The collection effectiveness of the tax is phenomenal; in the Consortium area 99.9 per cent of the tax owed was actually received.

Because drainage costs are pooled across the sixteen members of the Consortium, the drainage rate paid in one IDB, such as West of Ouse, may be equal to, higher than or lower than its actual drainage costs in any one year. The institutional skills that hold the Consortium together as a cooperative body show an understanding that, in the cycle of capital works and OM&M over the years, no one area will be singled out for special favour or disadvantage. The Consortium's Board of Management represents all the individual IDB s and the golden rule–"everyone gets his fair share", equal gain in the long term–is watchfully, if not systematically, monitored.

DRAINAGE NEEDS IN THE LOWER INDUS

The Left Bank Outfall Drain (LBOD, stage I) project in Pakistan provides a case study of a major investment to provide drainage services in an irrigated command area. It illustrates many of the physical and economic characteristics of drainage design and implementation and of the application of cost-benefit analysis to such investments. The problems of rising water tables resulting in waterlogging and salinization of soil are nothing new. About 4,000 years ago, the Indus river civilization that had flourished for 1,000 years at Mohênjodaro in what is now Sindh province came to an abrupt end. Some archaeologists have speculated that prolonged irrigation without drainage had caused the water table to rise. Floods and soil salinity then destroyed the agriculture on which the city depended. Other ancient civilizations, apart from those in the Indus valley, were also afflicted, for example, those that grew up on the flood plains of the River Tigris and River Euphrates in what is now Iraq.

In Sindh a major programme of improvement and construction of inundation canals was undertaken in the latter half of the nineteenth century, but it was not until 1932 that barrage-commanded irrigation was introduced with the construction of the Sukkur barrage system. This commanded a gross area of some 5 million ha on the left and right banks of the lower Indus. Later two other barrages, Kotri and Gudu, completed the system as it is today. The LBOD stage I project addresses the drainage needs of the left bank of the Sukkur barrage command. This is almost entirely a perennially irrigated area supporting a kharif crop of mainly cotton and a rabi crop of mainly wheat. In 1932 it was known that drainage would eventually be required, but the deep water tables at that time meant that it was not initially needed. Thus the LBOD project might be regarded as a deferred part of the original investment in the irrigation system itself that commenced with construction of the Sukkur barrage in 1932!

The water table in irrigated areas in Sindh has risen by as much as 27 m in the past 30 years and over 3 million ha are waterlogged. Although rainfall averages only 125-150 mm/year, 50 mm may fall within 48 hours in summer storms. Gradients on the floodplain are low and after storms water may stand in the fields for weeks, creating a public health hazard and damaging crops. Salinity, which affects over 25 per cent of the farmland, occurs when the high water table brings salts into the evaporation zone of the soil profile. Farmers dare not allow their land to lie fallow for fear that salt will encroach and in parts of Sindh "islands" of green fields are surrounded by a wasteland of abandoned land.

In this environment, continued diversion of Indus water without drainage will inevitably result in waterlogging and salinity and this in turn will lead to depressed crop yields and further loss of cultivable land. If water is added without drainage, the water table will rise and salt will accumulate if that water is saline. On average 1 ton of salt is added to every irrigated acre in Sindh each year. With summer temperatures well in excess of 40°C, water is drawn up through soil capillaries and evaporated within the crop rootzone, leaving behind the salt. Before 1932 all of the stage I project area had a water table less than 3 m deep; 50 years later 75 per cent of the area had a water table less than 1.5 m deep and a further 20 per cent less than 2.5 m deep.

Seasonal variations in water table levels are dependent on rainfall, whereas the degree of soil salinization depends on water table depth and irrigation applications. Water moves salts either upwards through capillaries or downwards by deep percolation. The closer the water table is to the surface the higher the evaporation, while for deep water tables capillary transfer and evaporation decrease rapidly.

Water table depth tends to stabilize where a balance with evaporation is reached. With no drainage and projected future water supplies, this would have been at about 1 m deep. Better water availability as well as seepage lead to waterlogging being concentrated along major canals, while tail-end areas usually experience water shortages and typically have deeper water tables.

The salinization process can be rapidly reversed by irrigation. Although irrigation is responsible for the long-term build-up of salinity, it is nevertheless sufficiently salt free for irrigation applications. With sufficient application, downward percolation will remove not only the delivered salt but also salt accumulated by capillary evaporation. Thus under intensive irrigation salinity can be controlled as long as drainage maintains the water table at a safe depth, or keeps the groundwater moving to prevent salt build-up.

For fallow or uncultivated land with a water table near the surface, water movement is upward and salt accumulates in the soil profile. For such an area adjoining irrigated land "dry drainage" may also occur. This is when irrigation water percolates to the water table, moves laterally and is evaporated up through the soil of the uncultivated land, thus causing rapid salinization. The two functions of drainage, water table control and removal of flood water after rainfall, are also not independent of each other because of the additional capacity of the soil to absorb storm water when the water table is lowered.

In designing drainage the depth at which the water table is to be maintained is inevitably a compromise between what is desirable and what is essential and affordable. The deeper it is the better the drainage, the better the ability to cope with normal fluctuations and the more storage is available to accommodate rainfall. Data on the relationship between crop yield and water table depth are scarce, but field studies in Sindh resulted in relationships being established for the principal crops. Deeper rooting crops such as cotton need about 1.8 m of unsaturated soil for full root development. It is also susceptible to damage from flooding. As noted above, the salinization of soil through evaporation from the capillary fringe is also substantially less at greater water table depths. Negative aspects are that some crops draw on the capillary fringe under conditions of water supply shortage. Moreover, in areas such as rural Pakistan hand pumps may be the main source of domestic water supply.

The LBOD project is designed to stabilize the water table at 2 metres or more, halt salinization and remove storm flooding more quickly. With drainage assured, farmers should be able to increase irrigation, achieve higher yields, reintroduce fallow into the crop rotation and face lower risks. Despite this apparent need for drainage its economic viability must still be assessed. Economic viability depends on being able to show that a greater planted area and/or higher yield of crops can be achieved after the land has been drained and has a sufficient incremental value of crops to pay for the costs involved.

There are other benefits to the infrastructure but these are relatively small and/or difficult to quantify, for example, improved road communication via drain embankments, reduced subsidence for existing roads and buildings and secondary economic benefits in the rural economy. Cost-benefit analysis for the LBOD project showed that because of its high capital and recurrent cost drainage alone would show an inadequate return on investment.

Rates of return for drainage investment alone in irrigation schemes are always likely to be low because the primary benefits–losses in productivity avoided–materialize gradually over a long time period. The economic viability of the project thus hinges on increased yield from a larger area of crops gained by rehabilitation and improved management of the water distribution system and of on-farm irrigation methods.

The planned drainage system has been described as a mirror image of the irrigation system. Stage I is the first phase of the project to be implemented, covering over 0.5 million ha of the left bank command and necessarily including construction of the large spinal drain which will take drained saline water to the sea. The infrastructure consists of the installation of drainage tubewells to extract groundwater hence lowering the water table and a network of surface collector drains that feed into the spinal drain. These will also act as interceptor drains to speed the removal of storm floods. A small area will be drained through the installation of sub-surface tile drains because the aquifer there is unsuited to tubewells.

Horizontal drainage in the form of tile drains was found to be substantially more expensive than vertical drainage using tube wells: combined capital and recurrent costs in terms of present value per hectare over the life of the scheme were approximately ten times greater. It is also only practicable to lay tile drains from about 1.8 to 2.5 m, as beyond the latter figure they become much more expensive. Under irrigated land the water may rise to 1.5 m at the mid-point between tiles but in fallow conditions will lie at an average of 2 m.

The water to be drained in the area is mostly highly saline; effluent from tile drainage can vary from 3,000 to about 10,000 ppm dissolved salts and that from tubewells can be twice as saline. A major design problem was to remove the water without affecting good agricultural land, but disposal is difficult-the choices being the Thar Desert River Indus, the Rann of Kutch, or the Arabian Sea.

Disposal into the first two was ruled out because of cost and hydraulic and environmental considerations. Outfall into the Rann of Kutch (a low-lying area of salt marshes and tidal lagoons) or the Arabian Sea was technically feasible, but the latter has the better hydraulic performance and environmental impact. In addition, it isolates the discharge from Indian territory. This latter point illustrates the possible political difficulties of disposing of drainage water, a mirror image to the difficulties of sharing river waters that cross international boundaries.

Drainage should halt the loss of cultivable land and sustain yields, but additional irrigation supplies are necessary for significantly increased yields and expansion of crop areas. As noted these gains are critical to the economic viability of the project. Additional irrigation supplies will come from measures taken within the project area and from sources outside. In the project area one of the best methods of providing the farmer with more water is to avoid wasting it in the first place. This is the function of the on-farm water management (OFWM) component that encourages the realignment and partial lining of watercourses (final distribution channels managed by farmers) to reduce seepage losses and also assists in precision land levelling to improve the efficiency of watering. A free design and supervision service plus a subsidy for 75 per cent of the cost of materials was made available to farmers for watercourse improvement. The objective during LBOD Stage I was that half the watercourses in the project area should be improved.

Along major canal lines, scavenger tubewells capable of separately pumping fresh (non-saline) groundwater from a lens that may overlay more saline deeper water and interceptor drains that run parallel to canals, recover some of the inevitable seepage losses for irrigation use.

Outside the project area Chotiari lakes are being used to provide local storage to regulate supplies while the Rohri and Nara main canal systems have been remodelled to carry additional supplies from Tarbela dam and (if built) the proposed Kalabagh dam. Together these inputs were predicted to result in a 50 per cent increase in water availability and a 40 per cent increase in cropped area, the difference reflecting the present degree of underwatering or water spreading. Note that where farmers are not badly affected by high water tables they tend to apply less than the recommended amount of water to crops, spreading it to cultivate a larger area. Although this reduces yields per unit area, it is the optimal strategy for profit maximization as it maximizes returns per unit of water, which is scarce compared with land. The practice does run the risk of not meeting crop needs if later deliveries fail, or not meeting leaching requirements and thus contributing to soil salinization.

The LBOD project concept described above was subject to intensive project preparation for more than 20 years. The definitive Lower Indus project study made the first proposals for a drainage project in 1966. The first full feasibility report for LBOD was produced in 1969, followed by a second project planning document in 1972. In 1975 the government of Pakistan began construction of the spinal drain using its own financial resources, but as a result of rising costs the project was revised and reduced in scope. In 1980 a study was undertaken with technical assistance provided by the UK's Overseas Development Administration that identified the LBOD component projects, established their priorities and recommended an overall implementation plan.

By 1982 when the World Bank took an active interest in the project some 100 miles (160 km) of the spinal drain had been constructed including an outfall into the Rann of Kutch. In 1982 the World Bank, concerned that the project had moved from a multi-purpose drainage project to a single-minded determination to construct the spinal drain–something they suspected would not be viable on its own–engaged consultants to carry out a fully detailed project preparation study for Stage I of the LBOD programme. Although this long history of planning studies and reports was not particularly planned or coordinated, it does mean that LBOD Stage I has evolved through a series of investigations, appraisals and processes of refinement during which it has been subjected to extensive professional scrutiny. This partly reflects the scale and importance of the project (to drain over 0.5 million ha at a predicted cost of US$636 million) but it is also an example, if a lengthy one, of how project preparation and appraisal in the water sector is likely to be an iterative process. The process may be further complicated by the complex negotiations and arrangements required to secure the financing of such a large project.

For the economic and financial cost-benefit analysis of the project the major cost items were the civil engineering costs of installing the drains and tube wells and their identification was thus straightforward. Identifying the benefits was more difficult–a characteristic problem for investments in drainage for which a large proportion of the benefits may derive from predictions of future losses in productivity that will be avoided. As described above the general contention is that the LBOD project will increase yields and cropped area over approximately 0.5 million ha of command by lowering water tables and salinity levels, saving or recovering water lost from the irrigation system, providing supplementary irrigation water and reducing flood damage.

A shift toward more diversified high-value crops, such as vegetables and orchard fruits, was also predicted. Incremental benefits depend on higher production in the "with" than in the "without" project scenario, depending very much on changes in land use. Without the project, rising water tables caused the intensification of cropping by farmers on their best land, as fallow areas were reduced and land abandonment increased. It was projected that further increases in irrigation supply would initially increase the cropped area but at the expense of non-cropped area, reducing the dry drainage effect and allowing the water table to continue rising. This would force the farmer to intensify further and all fallow land would potentially be lost.

Eventually the water table might rise to a level that inhibits cropping and cropped area would begin to reduce. Any further increases in supply would then cause wholesale abandonment of land. At the time of the project preparation study some areas had reached this point with pockets of cultivated land surviving because a saline waste surrounded them, providing the dry drainage that kept the water table just deep enough for cropping.

An initial study of marketing and pricing of agricultural products led to the conclusion that existing marketing, processing and pricing arrangements for the project's agricultural inputs and outputs were satisfactory and not likely to impose constraints on achieving projected production and farmer income levels with the project. High-yielding varieties and fertilizers had already been widely adopted and it was expected that increased water supplies would soon be taken up both to increase applications to existing cropping and to extend the crop area. The value of surface drainage was also clearly understood by farmers, who commonly ran off stormwater to the few drains that existed or to nearby depressions.

Less well known was their attitude to subsoil drainage, although there were some indications that the need was realized. Overall it was assumed that the farming community would respond well to the provisions of the project and the expected benefits would thus be generated. However, almost completely lacking from planning and the early stages of implementation was any participation by farmers in the project design or even an effective communication strategy to explain what was intended.

To date it is difficult to evaluate the success and impact of the LBOD project. Delays in implementation meant that drainage tubewells only began to operate over a significant part of the project area during the late 1990s. Operation and maintenance problems also inhibited the full and continuous operation of those works completed. Where operation has been sufficiently continuous there is localized evidence that reversing waterlogging and salinization has improved the sustainability of irrigated agriculture, while the change from dry drainage to drainage with disposal has enabled additional irrigation water to be used effectively and the area of abandoned land to be reduced.

These improvements have had a positive, though modest, impact on land values, agricultural production and rural employment. Evidence of yield improvements attributable to the project is much more limited. Many farmers still fail to invest sufficiently in the inputs needed to raise yields, in part because of the on-going failure by government agencies to deliver both irrigation and drainage services that are adequate and reliable. Other important factors have been poor price incentives, particularly for wheat during the 1990s and poor performance in the delivery of other services such as credit, certified seed and extension. An important lesson from this case is that drainage and irrigation delivery improvements alone will not necessarily improve the performance of agriculture if farmers also face other constraints.

It is also clear that effective strategies for operation and maintenance and for cost recovery are critical. This is difficult because for drainage there are additional problems to the normal difficulties of revenue generation from low-income farmers through irrigation charges. Farmers may regard drainage as not directly productive and that it should be a part of national infrastructure that is a government responsibility like roads. Indeed they may expect compensation for land lost to surface drains rather than levying of charges. Those downstream may claim that it is salt disposal upstream that is creating their need for reclamation and drainage. Upstream farmers are likely to be reluctant to accept a share of downstream drainage costs even though, as in the case of the LBOD, they are dependent on downstream maintenance of the surface disposal network.

Current drainage problems were also created by past decisions to defer drainage investment. The costs avoided have accrued as a benefit to either government or farmers who have cultivated without drainage. Should current farmers now pay the full costs of drainage or should government pay for its previous short-sighted investment decisions and deficient regulatory policies? How should such choices be weighed in the context of the other subsidies or taxes applied to other farm inputs and outputs?

Drainage can often be viewed as closely approximating the economic characteristics of a public good, i.e., non-excludability and non-rivalness. There is a joint or collective demand for drainage and if regional or area drainage is provided no farmers (or other residents) can be excluded from benefiting even if they pay no drainage charges. Once water tables are lowered across the area non-rivalness applies, as enjoyment of the benefits by one does not consume or diminish the benefits for others. Assuming farmers can afford the investment needed, agricultural drainage can be a private good if farmers can drain their own land and either dispose of the drainage water on their land or reuse it for irrigation.

Neighbours may still receive a positive externality from the lowered water table, or a negative one if disposal causes environmental damage. However, once collective disposal systems are needed public good arguments again apply. For the LBOD project the need to get rid of saline groundwater creates the requirement for a regional infrastructure of surface drains. Even for these excludability would be difficult and costly to enforce and non-rivalness applies as long as capacity is sufficient to accept the disposals from farmers.

In schemes such as the LBOD neither the benefits nor the infrastructure are evenly distributed, yet all components must be well maintained and operated for all to benefit. The costs of drainage cannot easily be subdivided and allocated, particularly where farms are small and the hydrological boundaries for the effects of tubewell drainage hard to determine. Similarly it is not easy to see a system of subdividing the benefits into purchasable units that can be competitively sold separately to different individuals. In other words, property rights–the basis of all markets–cannot easily be established for regional drainage. Here then is the drainage challenge. Essential for sustainable irrigation in many areas, economic theory suggests that public provision is necessary (at least for major elements of the infrastructure), but experience as exemplified by the LBOD suggests that developing countries have overextended public-sector commitments exacerbated by low efficiency and accountability of public agencies.

It is no mean challenge. For example, combined irrigation and drainage operation and maintenance costs for the LBOD scheme may be ten times greater than those for irrigation alone. In fact, farmers have been paying less than half of the irrigation costs. If pre-project predicted levels of benefits are achieved farmers should be able to pay such costs from incremental income, but as noted actual achievements may be rather less. Not adequately addressed during planning or early implementation, these issues are being tackled in the LBOD scheme through a number of pilot initiatives. Improved communication with the population of the project area and participation by farmers in the design and construction of on-farm works such as tube well disposal channels has brought improvements.

Farmer groups may be able to take responsibility for some operation and maintenance tasks but modalities for this are yet to be resolved. Use of private contractors and performance contracts for operation and maintenance has been trialled successfully, aiming to minimize the number of staff employed by government agencies while ensuring that equipment and workshops already supplied under the project are effectively used. Despite the importance of the scheme these and other unresolved funding and management issues threaten its success and sustainability.

PUBLIC GOOD AND THE CHRISTIAN EUCHARIST

Curiously, "public good" cannot be applied to any goods at all, at least in the sense of a good as an artefact, for clearly goods in that last sense *can* be privately appropriated. It is also important to recognize that the supply of a public good does diminish the benefits derived by consumers insofar as it draws down on resources that could have been applied to alternative goods and services. The competition for resources between guns and butter (national defence versus national consumption) is the starting point on the production possibility curve in introductory textbooks on economics. Nor should we confuse a public good with a public–sector good or service–the institutional ownership of the supplier has nothing whatsoever to do with the economic definition of a public good. Drainage, for example, is a public good whether or not it is supplied by a district board or a private company. It might be thought that a public good is one where the benefits to any single consumer cannot be identified. This too is false. For example, the treatment of polluted industrial wastewater prior to its discharge into a river is certainly the provision of a public good, but the change in the quality of river water abstracted by any single actor can certainly be carried out in great detail.

The theoretical economist's concept of a public good is a cure for which, unfortunately, there is no disease. What *matters* is that drainage and wastewater treatment once in supply can benefit a variety of catchment actors, the services commonly enjoyed in this way are non-exclusive and the measurement of the uptake of the service with its consequential advantages may be complex, open to dispute or just downright costly.

So the payment system for this service supply requires a design quite different from that of the bulk of goods and services in terms of both assessment and collection. It is this that makes the English and Welsh system of NRVs of such interest with respect to drainage. Within the last year or so Swedish institutions have begun financing studies of water resource management in catchments shared by more than one nation from the point of view of such management importantly constituting a public good. The resources applied would be better used to revive the theological discussion of transubstantiation.

Chapter 9

Analysis for Irrigation and Drainage

THE EVALUATION OF CAPITAL PROJECTS

Institutional economics is a policy-oriented discipline in which the significance of economic inquiry is a function of its relevance for problem-solving, including those dilemmas faced by governments. Virtually every central or regional government has some overview on how it sees its distinct economic sectors, such as that of water, developing over time. Within sectors, programmes will be elaborated and programmes require projects, the cutting edge of economic change.

The provision of government finance for a country's river-basin, water-supply and sanitation activities is always likely to face a budget constraint. It follows that, in choosing one set of initiatives demanding government resources, others inevitably have to be rejected or at least postponed. For example, Kinnersley has referred to the competition between irrigation projects and the provision of basic water supplies in some countries.

The question then arises, from the point of view of the country as a whole, of "What is the optimum combination of initiatives to select in any given year?" An entire branch of economics, known as social cost-benefit analysis (SCBA), can assist in answering this type of question. SCBA concerns itself with the evaluation of capital projects. There is no assumption that the technique is a substitute for politics; rather, it interacts with politics. The literature on social cost-benefit analysis is enormous. It then continues with issues specific to SCBA that the economist must confront: the choice of the discount rate; the use of shadow prices; the exclusion of transfer entries; and various estimation problems. Thereafter, the ground shifts to consider whether project analysis can have any relevance to demand management.

PROJECT DEFINITION

A project's characteristics are that it usually has a well-defined geographical location, it is an activity with a specific starting point at which time investment costs are incurred and during the period of the project's life a series of outputs are produced. The lining of a set of irrigation canals is an example. Gittinger notes: "Usually [the project] is a unique activity noticeably different from preceding, similar investments and it is likely to be different from succeeding ones, not a routine segment of an ongoing programme." The project cycle can be classified in alternative ways. Here is an example that highlights the process as an economic activity:

- *Identification*: In this stage the project is conceived and its broad outlines agreed. This might come about as a result of a sector survey of broader scope.
- *Preparation and analysis*: A more detailed homing-in on project definition occurs with feasibility studies of alternative approaches, covering economic, financial, technical and organizational aspects.
- Ex ante *evaluation*: The project is evaluated in economic terms prior to its launch, on the basis of forecast expenditure and income on capital and current account. *Ex ante* social cost-benefit analysis is appropriate at this stage.
- *Implementation*: This breaks down into the gestation period, when the bulk of the investment costs are incurred and subsequently the working period of the project until it comes to an end.
- Ex post *evaluation*: This takes place at the end of the gestation period and at any point during the project's productive life or at its end. In the last case, all the data are retrospective, in contrast with the forecast and prospective intelligence of *ex ante* evaluation. The value of *ex post* appraisal is in developing subsequent strategies, programmes and projects on the basis of the wisdom of hindsight; it also can expose systematic errors in *ex ante* evaluation.

The general objectives of evaluation in economic terms are to test whether the returns on projects exceed their costs, to rank projects with respect to their economic efficiency and to help policy-makers choose a set of projects, within the financial constraints facing them, which contribute most effectively to the country's economic development in a manner consistent with its social and environmental values. In the context of this book, SCBA is a technique primarily intended to raise the efficiency of government expenditure in support of water infrastructure programmes, particularly in respect of irrigation and drainage in agriculture.

In its preparation and analysis phase, the project cycle should include an appraisal of the impact of an investment proposal on specific actors, such as the inhabitants of an area to be flooded for dam construction purposes, the civil engineering industry, regional government, farmers and manufacturing firms. This is vital to understand who is likely to benefit or to lose from the project and what their anticipated behavioural response may be. But the method for the evaluation of capital projects concerns itself primarily with the costs and benefits of the activity from the point of view of the nation as a whole. For this reason, project costs and returns are considered in terms of the real economy, i.e., the flow of real resources used up in order to produce flows of real outputs or services. This holistic rather than partial approach is why the term *social* cost-benefit analysis is used.

THE NET PRESENT VALUE APPROACH

A starting point for understanding the technique of *ex ante* SCBA is to see it as a means to judge whether any single project will (or will not) make a positive contribution to a country's gross domestic product (GDP). This requires that–over the full life of the project, i.e., the gestation period plus the working period–one identifies the annual real flow of goods and services produced *with* the project in place, in comparison with the real flow of output *without* the project. This will be called the incremental real flow. At the same time we identify the incremental annual real flow of resources used in order to produce that output.

These identified physical quantities of outputs and inputs, in the form of goods and services, can then be given a common standard of measurement, in terms of their market prices. A spreadsheet may be drawn up of the annual value of the incremental real output in row:

- The value of the incremental real costs in row.
- These two rows are hereafter referred to as the project's benefit and cost flows. The difference between benefit and cost in each year is the net benefit flow and is recorded in row.
- All this has been done for a water project deemed to have a full life of 25 years, starting at the beginning of year 0 and terminating at the end of year 24.

Where the project has clear external effects in GDP terms that the producer neither pays for nor receives payment for and where these are quantifiable in price terms, such positive or negative externalities should also be included in the benefit, cost and net benefit flows. Kinnersley has pointed to the frequency of negative externalities from water projects:

In economic terms, the most common hazard is the way in which rivers spread damage caused at a particular point. Activities that pollute the river or divert part of its flow usually affect water quality or flow for some distance down stream. The water may thus be made unsuitable for other uses down stream, but often without would-be users down stream being able to trace who or what caused the pollution and thereby claim compensation. Thus, the polluting upstream activity, assuming it is a business, may be said to be escaping costs of waste disposal and preventive measures against pollution which it should normally have to provide for and reflect in its selling prices.

In the early period of a water infrastructure project, the net benefit flow is likely to be negative, when capital account outlays fail to be matched by the stream of benefits. Thereafter, net benefits may remain positive through to the end of the project's life, although major rehabilitation outlays in the mid-years could modify this general rule. In all cases we would expect the identification and valuation of benefits to be even more difficult than those of costs.

An important issue arises immediately. Should the market prices we are using for the valuation of the incremental real flows be constant or out-turn prices? Constant prices are those ruling in one specific year and are applied to the outcomes of both that year and all other years considered in the analysis of the project. Out-turn prices are the ruling level of prices in each separate year and, of course, vary over time. The standard practice in the field is to use constant (usually base year) prices. The argument is that a general price inflation does not, in comparison with zero inflation, indicate any difference in the real value to society of outputs or inputs. This rule notwithstanding, the planning of a project's financing should consider the likely level of general price inflation, so that an adequate budget is obtained to cover the costs of the gestation period.

However, if there is a strong expectation of a *relative* shift in the prices of some outputs or inputs, quite aside from any general movement in the price index, these would be reflected in the calculation of the net benefit stream. Here the assumption is that relative change implies a shift in the value that society, through the market, places on those commodities.

A second issue that deserves to be addressed at an early stage concerns the mode of project finance. Should SCBA as a general practice embody in its calculations the sources of the money that will fund the project during its gestation period? Gittinger argues persuasively that this should not be the case. He suggests that one assumes all financing for a project comes from domestic sources and that all its returns go to domestic residents. This is consistent with the general objective of SCBA defined in terms of projects' contributions to GDP. Project analysts almost universally accept this convention of splitting the evaluation of individual projects from their specific financing sources. It is perfectly consistent with identification of possible financing sources in the preparation and analysis phase of the project cycle that precedes or runs alongside *ex ante* appraisal. The convention also neatly side-steps the fungibility enigma.

By this stage, we should have completed a spreadsheet. Let us suppose that N is positive, i.e., the total value of the stream of incremental real output *(B)* exceeds the total value of the flow of incremental real costs (*C*). In such a case, is it correct to assume that the project should be approved? Can SCBA really be that easy?

Unfortunately it is not, for two distinct reasons. First, such a procedure would ignore the timing of the benefit and cost flows. Second, such a procedure would provide no basis for choosing between projects, either where two (or more) are mutually exclusive for technical reasons or where a budget constraint imposes a limit on the total number of projects that can be financed-the starting point. The term *project interdependence* will be used to refer to either of these two situations.

With respect to timing, in project evaluation it is universally accepted that each annual benefit and each annual cost entry should have its value reduced, i.e., it should be discounted, the later it occurs. The mathematics of discounting is set out below. The accounting conventions are that the total project period is divided into years, that rates of growth or contraction are expressed in annual compound rates and that all receipts and payments in each year occur at the beginning of that year.

Suppose US$1 is received at the beginning of year 0 and that in each successive year the sum received increases at a constant compound rate of 5 per cent. Then, after 5 years this series of payments in dollars and cents is as set out.

Thus, the entry for each receipt in any given year t is equal to:

$$1.00(1.05)^t$$

More generally, for any given proportionate rate of increase i, expressed as a decimal, we can write the receipt as:

$$1.00(1+i)^t$$

This is a compound growth series. But when we discount the future, we are suggesting not growth into the future but, so to speak, contraction back from the future. We argue that an entry t years after the present (the start of year 0) needs to be reduced proportionately each year to calculate its present value.

$$\frac{1.00}{(1+d)^t}$$

The decision to discount each and every entry after year 0 should be carried out, as equation above indicates, through dividing it by $(1+d)^t$ where, again, d is the selected discount rate (such as 5 per cent) and t is the year in which the entry appears.

These individual discounted values of the benefit, cost and net benefit streams are added up and recorded. They will be written here as B^*, C^*

and N^*. N^* is known as the net present value (NPV) of the project. Where the NPV is positive, the project can be approved on economic grounds, provided project interdependence does not exist.

The apparently laborious task of dividing dozens of cell entries by $(1+d)^t$ can be simplified with a calculator, with a published discounting table, or with an NPV function available through standard spreadsheet computer software. In the approach above, all costs and benefits in the first year are undiscounted, since they are assumed to occur at the beginning of year 0. [Note that $(1+d)^0=1$.] Alternatively the first year of the project is counted as year all entries are deemed to have been incurred at the end of each year and, once again, we have the convenient expression $(1+d)^t$ to use to discount, where t is the project year in which the entry falls. Here, first-year receipts and outlays *are* discounted.

Next it is necessary to consider the difficulty that, in common experience, not all project proposals are appropriately considered in their own right–as independent of each other. In the first place, on technical grounds two (or more) proposals may be mutually exclusive, most obviously when each is an alternative means by which a water infrastructure investment can secure a given policy objective. In the case of mutually exclusive projects, the solution is simple: calculate the NPV of each option and choose the proposal with the highest NPV.

In the second place, projects may not be appropriately considered as independent because a single programme budget is available for their financing and the budget is insufficient to fund them all. The question is then posed: "On economic grounds what is the optimum subset of proposals to which finance should be provided?" In this case, for every proposal (including all mutually exclusive projects) we calculate the net benefit-investment ratio (NBIR). This is simply done by dividing the full project life into two: the gestation period, when the net benefit flow is consistently negative and the working period, when the net benefit is, in most years, positive. The NPV for each period is separately calculated and that of the working period divided by that of the gestation period. The negative sign of the denominator is ignored; absolute values are to be used here. All projects are then ranked from top to bottom, using the net benefit-investment ratio. Lower ranking mutually exclusive projects are eliminated and then all the remaining projects are approved off the top until the budget constraint blocks any further approvals.

What we have done in the second case-that of financially interdependent choices–is to consider the key constraint to be the budget funds available during the gestation period and then to rank all projects in terms of the productivity (in NPV terms) of these funds. This maximizes the total NPV generated by the use of the funds available. The net benefit-investment ratio can also be considered as a means not of excluding projects but of ranking them in terms of their starting date.

When applied to irrigation and drainage projects, the net benefit-investment ratio is a valuable output-input ratio and we shall refer to it as E_{10}, yet another measure of irrigation water efficiency.

Discounting the Future

Up to this point, the NPV approach to the evaluation of capital projects has been presented in terms of its most basic elements. But if SCBA in this form is to be an acceptable technique for project decision-making, some extremely important areas of debate must be introduced. These are classified under four headings: discounting, shadow prices, transfer entries and estimation errors. It was argued that the timing of the benefits and costs of a project should be handled in the evaluation process by reducing these values in proportion to the time delay before the benefit or the cost is registered. The method for doing this was described as the division of each value by $(1+d)^t$, where d is the selected discount rate and t is the year in which the entry appears. But no reason was given as to why the timing of benefits and costs is significant, nor why they should be handled by means of a discounting process.

The justification for this approach falls not within the field of economics but of psychology. There seems to be an almost universal human propensity to regard a real benefit offered now as superior to the same real benefit offered later in time. Human life is beset with risks of injury, loss and death. That has always been the human condition and is doubtless the origin of a rather rational desire to receive now, when the probability of receipt is greatest, rather than later, when such probability is perceived to be less. This mind-set is an example not of human frailty but of sweet wisdom distilled from the bitter lessons of experience.

Project evaluation, rightly in my view, recognizes this psychological propensity by the use of the discounting technique in respect of project outputs. This still leaves us with the difficult question: "At what level should the time rate of discount be set?" Since the evaluation of capital projects is here placed within a national framework, setting out from the concept of the GDP, one is clearly seeking a value of d that is appropriate to society as a whole and so one refers to this rate as the social time rate of discount (STRD).

To grasp the reasoning here, it may be of some help to draw a parallel with the physicist's concept of the half-life of radioactive material. In applying the arithmetic of discounting to handle the psychology of risk avoidance, we are suggesting that a real output (or cost) available now with a value of US$1.00 has proportionately lower value today if it is forecast to be available only in a year's time and a still lower proportionate value for the following year. And so on. Thus today's value of that projected real output contracts or decays at a constant rate the later the time at which the benefit is forecast to be available, rather like the radioactive strength of some elements decays at a constant proportionate rate over time. For any given value of a real output (or cost) at the beginning of year 0 and for any specific measure of the social time rate of discount, it is easy to say how many years will elapse before that original value is deemed to have decayed by one-half. This number of years we shall call the half-life of the value of incremental real output implicit in any measure of the STRD.

The definition of the half-life associated with a specific social time rate of discount needs careful specification. For any real benefit valued today at US$1 if it is available today, but valued today at only US$0.50 if it is forecast to be available in *h* years time, then *h* is the half-life. The half-life of the value of output, cost or net benefit is given for every integer value of the STRD between 1 per cent and 15 per cent. The results are dramatic. At a value of 1 per cent, the half-life is as long as 70 years. But at 15 per cent, the half-life is only 5 years. The absolute value of the half-life falls substantially up to an STRD of about 5 or 6 per cent, where it averages close to 13 years.

This figure may help decision-makers to set the appropriate rate for project evaluation in their own countries. In my view the whole thrust of discussions on sustainability in recent years suggests that high values of the STRD are inappropriate, as they produce short half-lives–they markedly devalue the medium- and long-term effects of project investment decisions. This point of view has been strongly urged by Lipton (1992) in the context of the high real rates of interest that were current in the 1980s and 1990s. A benchmark STRD of 5 per cent per year may be widely appropriate, which would give a half-life for net benefit entries of about 14 years. An alternative to the NPV approach is used by organizations such as the World Bank and most other international financing agencies, as Gittinger reports. This calculates the internal rate of return (IRR) of a project, i.e., the discount rate at which the discounted value of incremental real output precisely equals the discounted value of incremental real costs. A cut-off rate is then set and all independent projects above this value are deemed worthy of approval. Note that when this IRR is based on the inputs and outputs in financial terms it is called the financial rate of return (FRR); when based on inputs and outputs in economic terms we have the economic rate of return (ERR). There are problems with the IRR approach. First, the cut-off rate is said to represent the opportunity cost of capital in a country, but in fact no one knows what this is. Moreover, the concept of the opportunity cost of capital rests on the assumption that its use would lead to the approval of all possible investments scoring above the cut-off rate. Yet such an outcome is entirely inappropriate, since choice on that basis ignores the huge importance of environmental and social effects in correct decision-making.

Second, the ERR cut-off rate is used as a crude ranker of projects, but it is recognized that the ERR should not be used to rank projects at all. Third, when constant prices are being used, chosen cut-off rates are typically in the range 8-15 per cent.

If the appropriate STRD is much lower, as I have suggested, then the use of such a high cut-off rate for the ERR introduces a bias against sustainability, favouring projects with lower (but earlier) undiscounted returns.

The advantages of using a net present value approach based on the social time rate of discount are that:

- It forces decision-makers to make choices about the appropriate social time discount rate in their society.
- Once this is set, no unjustifiable bias in favour of the short term takes place.
- It gives a practical ranking for interdependent projects, using the net benefit-investment ratio, which recognizes in a pragmatic way the opportunity cost of capital funds.

The use of the IRR at a high discount rate can have powerful effects on the routines of design. Stephen Allison writes:

> After I finished my Ph.D. I worked for Harza Engineers in Chicago and, in the course of the first six months, wrote an internal guide entitled "Designing for a 15% IRR". That was the Bank's (almost only) criterion in the mid-sixties. I made the point that if one did a preliminary feasibility study which yielded 13%, the thing to do was seek places where less equipment or construction materials could be used that were less durable and of lower capital cost. Early replacement, or higher O&M costs, would not "hurt" the project as much as higher capital outlay!

Before moving on, a few brief words are in order on SCBA's hard-worked cousin–social cost-effectiveness analysis (SCEA). Broadly speaking, whenever two or more projects achieve the same outcomes but in different ways, they are mutually exclusive on technical grounds. Project analysis limited to such a package of alternatives can be carried through merely by calculating which option has the lowest discounted social costs, i.e., which is the most cost-effective from the point of view of society as a whole. Winpenny for example, cites the case of irrigation canal lining in southern California in the use of SCEA.

SHADOW PRICES AND TRANSFER ENTRIES

Project evaluation should always do this for its first estimate of the NPV. However, there may be cases where the market price of an input or output is not deemed to be the best measure available to assess the project's contribution to GDP. This is usually where the market price of an input does not measure its opportunity cost adequately. The opportunity cost of a real input to the production of goods or services is the output value forgone elsewhere in the national economy as a result of that input's use for the project under appraisal.

Shadow Prices

Where a unit price other than the market price is used, so that the values of incremental outputs and costs are recalculated, we speak of *shadow prices*. In the text below are listed the principal situations where shadow prices are used. The topic is dealt with fully by Gittinger.

- In projects that use unskilled labour in a local labour market area where unemployment or underemployment is high, the shadow price for the wages of these workers should be set equal to zero. In the case of interdependent projects, the shadow price procedure will strengthen the ranking position of initiatives that are intensive in their use of unskilled labour. The same argument applies to semi-skilled (or even skilled) labour, provided un- or underemployment is high in their case also. In general, a zero shadow-wage for specific groups of workers will favour irrigation projects, if they are labour-intensive, in comparison with capital-intensive non-water projects. In their work in Pakistan Carruthers and Clark discovered that the effect of shadow-pricing the labour employed in drainage options was to favour hand-dug open drains over tile drains.
- The pricing of land will usually be at its market price, or an estimate of this based on its periodic rent. Where this price is believed to be distorted upwards as a result of local monopoly power, a lower price is appropriate based on land values derived from active rental markets. Dams with large storage capacities take up vast quantities of land. However, for each million cubic metres stored, small dams probably use even more.
- Where the project's raw materials and manufactured inputs are purchased from industries operating at excess capacity, they should be valued only at their marginal cost, not their market price.
- Projects producing tradeable goods (i.e., outputs that are actual or potential exports) must now be considered, as well as schemes

using imported inputs. In many developing countries, the domestic currency is overvalued at its official rate. For example, the official rate of the dollar to the peso might be 1:1 rather than the market rate of 1:2. Whether the SCBA is carried out in dollars or pesos, overvaluation (which is often accompanied by tariffs on imports and subsidies to exports) tends to favour projects that are relatively intensive input importers and to disfavour those that are intensive output exporters. In such cases, SCBA would use a shadow rate of foreign exchange closer to the unofficial market rate. If a developing country or transition country's currency is overvalued, irrigation projects that stimulate food exports or which reduce agricultural imports would be strengthened by the use of a shadow price. However, infrastructure projects that are dependent on imports, for example, of construction machinery, steel and pump sets will find their NPV cut sharply through the shadow-pricing of foreign exchange.

- Where a project's outputs are large relative to the existing volume of domestic sales of those crops, the market price is likely to fall. Conventional practice is to shadow-price output half-way between the market price with and the market price without the project.

It would be absurd to expect the project analyst of any single investment to prepare from a blank sheet a set of appropriate shadow-prices. These should be set nationally and may be available from the Ministry of Finance.

Transfer Entries

In the calculation of the NPV, the objective is to identify real outputs and real costs and then to value these at market (or at shadow) prices. In the *financial* accounts of a project, where project cash-flows are the focus, certain transfer items usually appear. These include the capital sum of a loan, repayment of loan principal, payment of loan interest, receipt of government subsidy and payment of government tax. They represent only transfers of money between one actor and another, not the values of real outputs and inputs from the perspective of the national or regional economy's GDP. So, in estimation of the economic rate of return, in contrast to the financial rate of return, these items would not be included in the net benefit stream. Note that insurance payments are usually taken to represent a real cost, not a transfer payment.

Ex ante evaluation makes an estimate of the project's NPV through forecasting future events such as the length of the full project life, the real costs of producing real outputs, the future market prices of inputs and outputs and the future setting of shadow-prices. Inevitably, *ex ante* analysis makes errors of estimation. When the project is given its go-ahead, or is turned down, its NPV is a hazardous projection, not a fact.

Yet, *ex post* evaluation also faces substantial estimation problems, captured by Kundera in the wonderful phrase "the unbearable lightness of being". Let us suppose a major land drainage project is completed in 2002 in a farming area where low tonnage efficiency exists alongside extensive unemployment.

The justification for project approval had included positive reference to the expansion of employment in the local labour market area. In the year 2005 an *ex post* evaluation is carried out. Actual employment in that year is compared, favourably, with actual employment in the year 2002 and the argument is made that the original project approval was, on employment grounds at least, fully justified.

Unfortunately this commits the most elementary error of project analysis. For the incremental approach adopted into a *change* over time but to the *difference* in outcomes with and without the scheme. The employment difference between situations with and without the project is much smaller than the actual changes in employment, because in any case employment was on an upward trend, the project's employment effects are much weaker than the comparison of the years 2002 and 2005 indicates. The unbearable lightness of being is that we can never know, in *ex post* evaluation, what would have occurred without the project in place, because that did not happen. We cannot rewind history to discover what would have happened if the project had been refused funding. So, there is a curious symmetry between *ex ante* evaluation, which forecasts one unknowable future and *ex post* evaluation, which reflects on one unknowable past.

However, these evident difficulties should not be allowed to induce the paralysis of reason by the abandonment of analysis. Little and Mirrlees argue strongly that the net advantages of SCBA are considerable. Evaluation can be thought of as the partial elimination of error implicit in decisions made by hunch or on the grounds of apparent financial profitability. This suggests that NPV estimation procedures should focus on elements that seem likely to make a substantial difference to the outcome, rather than seeking greater accuracy in respect of trivia. Where powerful determinants of project success with a range of possible values are identified, sensitivity analysis can be used to test NPV response to different, plausible assumptions.

Just as important as the reduction of estimation error is the avoidance of estimation bias. In the assessment of project outcomes, specific irrigation actors–in the private or the public sphere–will always have a material interest in specific projects receiving approval and others being rejected.

A quite specific form of estimation bias, known as the McNamara effect, has unfortunately marked World Bank project appraisal for many years. When the supply of funds for projects is in full flood, pressures of an informal kind are placed on appraisers to inflate their calculation of the rate of return, so that more projects qualify for approval. Such manipulation can prove important for the internal promotion of the project analyst. Moreover, the analyst will have moved up, up and away should the project crash. Little and Mirrlees suggest that the McNamara effect is found not only in the World Bank but in much commercial bank lending to the Third World. In mitigation, a Bank source informs me that in addition to the rate of return it does use qualitative assessments where benefits are not readily quantifiable, such as multiplier effects in employment generation through water availability.

Even in the new millennium, the McNamara effect is not dead. Stephen Fidler in an article on James Wolfensohn, the President of the World Bank, writes: since [Wolfensohn] took over in 1995, there has been a complete change of senior management and turnover of close to 20 per cent among more junior ranks. Yet officials say some flaws from the McNamara era remain. Staff are still rewarded for pushing loans through its governing board, even if they have not thought much about the longer-term consequences to the country involved.

The quite extraordinary feature of World Bank project funding is that, although the ERR is always calculated prior to project approval and then re-estimated at the end of the gestation period using the project completion report, further *ex post* evaluation is rarely carried out. As a result, no substantial analysis by the World Bank exists on the success or failure in SCBA terms of the Bank's own projects over the years of their working lives The same Bank source already referred to above reports that *ex post* evaluation continues to be sparse in the new millennium.

The issue of the relation between *ex ante* and *ex post* evaluation has by now a long history as the quotation below indicates: During preparation of this book a request was made to an international agency for *ex post* evaluations of irrigation projects. Although they have assisted with planning several hundred irrigation projects over a number of years, they replied: "Our studies are mainly of a pre-investment nature, rather than post-evaluation of operating irrigation schemes and we do not therefore have relevant literature based on these projects." Clearly, they consider pre-investment and *ex-post* evaluation as independent and unrelated activities.

The same authors argued on the basis of their extensive experience in the 1970s that *ex ante* studies typically underestimated costs and overestimated rate of uptake of irrigation water, yields and product prices.

"Almost all studies recommended that construction should be undertaken and that the scheme was financially and economically viable".

THE MANAGEMENT OF DEMAND

The orientation has been to the supply side of irrigation and drainage. This made sense because the focus is project analysis and new abstraction investments, reservoirs and other irrigation infrastructures are all meat and drink to SCBA. Yet, in the housing and residential infrastructure field, Malpezzi has demonstrated that SCBA can also be applied to initiatives that are not necessarily projects in the terms set out by Gittinger.

This raises the possibility that the technique could also be deployed in appraising demand management activities. Identified six strands to such action: internal and external reuse, use technology, land-use planning, environmental education, water pricing and food import policies. Internal reuse-the supply of agricultural drainage as a base flow for irrigation–can certainly be regarded as a traditional project for SCBA purposes. The same is true for external reuse, where household and industrial wastewater is used as an irrigation base flow. Where new irrigation technology is brought in to increase tonnage efficiency by augmenting gross crop acreage, rather than to cut water consumption, once again SCBA can be applied.

But land-use planning, environmental education and water pricing are all directed at reducing the volume of water used for irrigation purposes. These three types of measure, even if they cannot be regarded as projects in the traditional sense, certainly incur social costs. Land-use planning incurs a wide variety of current and capital costs. Environmental education demands important current expenditures, such as developing teaching materials and conducting awareness-raising campaigns. Water pricing may require metering as well as the staff to calculate and secure payment for individualized bills. And of course there may be a fall in the value of agricultural output in comparison with the no-project situation.

So, we can get a handle on the costs of demand management in the case of land-use planning, environmental education and water pricing. It will be important to remember that, in their calculation, one should subtract any existing outgoings that are no longer necessary. For example, billing for metered water will eliminate the need to collect the fixed charge, so the net cost of the initiative will be lower than its gross cost.

Moreover, supply costs such as the electricity used for pumping purposes will also fall. But what about *benefits* in the cases where land-use planning, environmental education and water pricing succeed in reducing irrigation water use in comparison with the no-management alternative? There are two cases to consider here. In case 1 the objective of demand management is to reduce the abstraction of surface water and groundwater in order to protect the environment. So the benefits are these environmental gains. In case 2 the objective is to allocate more water to households, manufacturing, mining and other commercial or municipal needs. The benefits here are the increase in households' welfare with the water services they receive and the value-added increases in the other sectors.

Finally, it is interesting to note that whereas abstraction licensing is a supply-side activity, not a dimension of demand management, it is nevertheless open to SCBA in a manner similar to demand management initiatives. The costs of abstraction licensing, where it is carried out to reduce the appropriation of surface water and groundwater for irrigation, are the regulatory process itself and the consequential fall in farm output. The gains are either the environmental benefits of a parallel increase in the natural flows of water, or the returns to increased abstraction for urban use. In the UK Knox have taken a particular interest in the GIS-mapping of the financial benefits of sprinkler irrigation and the potential financial impact of restrictions on abstraction.

SEDIMENTATION REDUCTION ON LUZON

When irrigation canal networks are sourced by water flows diverted from a local river, the abstracted volumes may carry high sediment loads. The canal network is likely to have a limited sediment transport capacity and so sediment deposition takes place, raising canal bed levels. In most cases this does not lead to an immediate reduction in conveyance capacity. Operators simply allow the water level in the canal to rise. But eventually the canal's freeboard limit is reached and dangers of overtopping arise. Thereafter the sedimentation process begins to throttle the canal. As the rate of discharge falls, the hydraulic capacity of the channel to transport sediment diminishes and the rate of sedimentation *increases,*

Sediment deposition of course cuts back the irrigation network's capacity to supply water to farmers. The area that can be irrigated sufficiently to meet its net irrigation requirements then reduces. Water supply across the network becomes inequitable and unreliable, with the downstream sections suffering most. Yields fall. In turn the relationship between farmers and the irrigation agency deteriorates and the irrigation service fee recovery rate slips back.

It is true that maintenance gangs exist to remove sediment. When the silt deposition rate is low, this may work well. But, as suggested above, sediment loads may be high in the water diverted from a river and the rate of sedimentation from a given load may accelerate as the canal's hydraulic capacity declines. In this case the maintenance budget–often dominated by the costs of sediment removal–may be insufficient to retrieve the situation, particularly if service fee income drops.

Various management or project options exist for dealing with downward spirals of this type. They all aim at reducing the sediment loads entering and settling in the canal network. A relatively low-cost, rapid-impact solution is to build sediment exclusion structures located at the entrance to a canal network. This case study reviews a cost-benefit analysis of just such a project carried out by Chancellor and co-workers in the Philippines. The research was published from HR Wallingford in 1996.

The River Agno irrigation district is located on the island of Luzon in the Philippines. The river's catchment area is some 39,000 km^2, about half of which in 1993 was forest and about half open grassland. Steep slopes and geologically young rocks that weather and easily fracture characterize the higher parts of the basin.

The area is subject to earthquakes that trigger landslides, which in turn provide a massive supply of easily erodible sediments. The river's upper reaches also pass through the Baguio mining district where up to 30,000 tonnes of mine tailings are produced daily, of which substantial quantities end up in the river through overtopping or tailing dam failures. Annual precipitation is some 3,600 mm, falling mostly in July-December. High rainfall intensities in storms and typhoons cause a high rate of soil erosion in unprotected areas.

The district's irrigation system was constructed in the 1950s with a final design service area of 11,500 ha. A conventional intake on an outside bend of the river diverts water to a designed 38 m^3/s main canal off which run more than twenty lateral canals. The wet season flows of the river are very much larger than the area's net irrigation requirements and in those months the irrigation area is controlled only by the discharge capacity of the canal system. The wet season service area began to decline long ago and this was attributed mostly to canal sedimentation. By the early 1990s effective irrigation had been confined to the head reach and parts of the middle reach zones. Chancellor and her colleagues write:

> Large scale desilting works carried out under the World Bank funded Irrigation Operation Support Programme (IOSP) in 1989/90 did increase the irrigated "total desilting costs" as the annual with-project benefit. After 1993 it is assumed that the beneficial reduction in costs falls at a rate of 4 per cent per year. This is because of the rise in sediment load as the total volume of water admitted past the cross-regulator grows with the increased hydraulic capacity of the canal network.

Benefits from increased agricultural output are derived from estimates of the annual growth likely in the wet season irrigated area. The study data suggest that the irrigated area with project would expand in comparison with the without-project area at an annual rate equal to 4 per cent of the base year hectarage. In respect of yields, no projected growth per hectare is made although, in the opinion of the researchers, this is quite clearly another instance of a "conservative" view on project benefits. Finally, then, this part of the benefit calculation was based on multiplying the additional wet season irrigated area by the estimated average gross margin in 1991 of the existing farmed area. These gross margin data were collected by means of a short questionnaire applied to a sample of farmers along laterals chosen to represent conditions in the head, middle and tail reaches of the irrigation district.

Chancellor and co-workers evaluated the project using the economic rate of return and included a number of sensitivity variants. The results were in all cases most impressive. The ERR varied between 22 per cent for the worst-case scenario up to 297 per cent for the best-case scenario. Different assumptions about the appropriate social time rate of discount. The results are phenomenal. Only a combination of world revolution, the plague, earthquake and volcanic eruption could create doubts about the economic viability of this project.

These incautious remarks of mine should, perhaps, be counterbalanced by the wiser words of the research team:

> Recovery of irrigated area usually has a very large impact in a cost benefit analysis and sediment control structures can be justified by a relatively small increase in cropped area. However, predicting the magnitude of area or yield increases is not trivial. In the retrospective study carried out at Agno, increases in irrigated area used in the economic analysis were based on observed data. This is clearly impossible in a study carried out to design and justify sediment control structures. A simulation approach provides the best means of deriving the technical information needed, but the farmers' response to improved water supplies and the willingness of those in former marginal areas to pay an irrigation service fee has to be carefully judged. Area and yield increases predicted as a consequence of improved sediment control should thus be viewed with some scepticism and benefits should be treated conservatively.

Disappearing Water

Perhaps the most extraordinary book on water resources ever published is Marc Reisner's *Cadillac Desert: The American West and its Disappearing Water*. Begun in 1978 and finished in 1985, with an afterword written in 1992 for the revised edition, it provides an absorbing account of the development of dam projects in the western United States in the great era of their construction between the 1920s and the 1970s.

In doing so it gives an acute insight into the nature and relevance of the social cost-benefit analysis of irrigation projects in those years in states such as Washington, California, Arizona, Colourado, Nebraska, Wyoming and Idaho. In the twentieth century a quarter of a million dams were constructed in the United States, of which 50,000 were major works. A couple of thousand of these, in Reisner's words, were "really big dams". To understand this history it is necessary to grasp the driving motivation behind the projects, the directed will and purpose of individual men and women located within hierarchically structured institutions. There is no better starting point than the two huge, public-sector bodies: the Bureau of Reclamation and the US Army Corps of Engineers.

The Bureau's mission was to develop reservoirs that would promote the growth of irrigated agriculture. Not unreasonably the organization sought to carry through such projects in many thousands of cases. But there was also a social psychology of the individual actors within the Bureau-in particular the engineers, although not confined to them–that rivers exist primarily in order to provide sites for dams and that the free flow of water to the sea is an enormous waste. For them *the conservation of water meant its impoundment and abstraction.*

The US Army Corps of Engineers had far wider responsibilities than the Bureau. One of these was flood control and, in part, this took the form of dam construction. But if a dam was said to be required for flood control, there was nothing to prevent its use as a reservoir feeding an irrigation scheme. As a result the Corps and the Bureau became vicious competitors in the long race to identify and build on dam sites. The social psychology of professional staff, with respect to reservoir construction, appears to have been the same in the two institutions. A key legal difference was that, if a dam was built for flood control purposes, the irrigation water could be supplied free to farmers.

Capital expenditure on dams by the Bureau and the Corps was most immediately beneficial to the local and regional construction and engineering industries. Firms such as Bechtel and Morrison-Knudsen became giants as a result of their contracts for the Hoover Dam. The Marshall Ford Dam, a 1930s Bureau project on Texas's Colorado River a few miles from Austin, was constructed by two small-time, road-paving contractors named Herman and George Brown. Before it was half-built the emergency appropriation funds for construction ran out, threatening the brothers with financial calamity. Their skins were saved by a young, newly elected Democrat congressman.

As a result of his intervention, the project went ahead with a further

US$5 million budget. The brothers' firm, Brown and Root, became one of the largest construction companies in the world. The youthful politician, Lyndon Baines Johnson, was to become President of the United States. Of course it is farmers who are presumed to be the main beneficiaries of irrigation projects. In the history of the American West the first irrigation farmers were Mormon families wishing to build a land or a place of their own–a search for ontological security. The Mormons enjoyed a close-knit society with a common faith but one that had a history of persecution. It was the 1862 Homestead Act that settled on 160 acres (64.8 ha)–"a quarter section"–as the ideal area for a Jeffersonian utopia of small farms.

The idea was to carve millions of quarter sections out of the public domain, sell them cheaply to restless Americans and arriving immigrants and, by letting them try to scratch a living out of them, develop the nation's resources and build up its character. The settlers had allies in the provincial newspapers and the railroad companies. Small farmers were hugely subsidized not only by the Corps of Engineers but also by the Bureau of Reclamation. Federal irrigation projects were financed at the Federal, not the State, level. *No interest was paid on these loans*. Farmers were required to pay back only the principal of the loan.

Meanwhile the provisions of the 1862 Homestead Act and succeeding legislation were becoming obsolete. Farming became an activity not just for families but for corporations such as Exxon, Getty Oil, the Prudential Insurance Company and Tenneco. Farm sizes in the agribusiness sector became very large indeed. It appears too that farming families and agribusiness in many cases did not even repay fully the capital cost of the irrigation infrastructure. According to a Natural Resources Defence Council report of 1985 on the Central Valley Project (CVP) of California, "the repayment of capital costs of the CVP is likely to be *zero* by the time most of the water contracts expire in the 1990s".

We have seen that the economic and institutional interests committed to the development of irrigated agriculture in the American West spanned the Bureau of Reclamation, the Corps of Engineers, engineering and construction companies, farming families, agribusinesses and many other associated regional economic groups. These bodies then exercised their economic and political influence with the United States Senate and the House of Representatives. Between the 1920s and the 1970s, thousands of elected politicians in Washington, DC, used their considerable skills and power to promote state dam projects for flood control, hydropower, irrigation and water supply to urban areas.

This hydraulic imperative took the form of the pork barrel. Any single project won by a senator, for example, would have its costs met by the federal taxpayer. Politicians supporting a project in a state other than their own believed that senators and congressmen from that state would in time return the favour. Moreover in getting projects under way, the politician could play off the Bureau against the Corps of Engineers in seeking arguments for its engineering and economic feasibility. "Every Senator... wanted a project in his state; every Congressman wanted one in his district; they didn't care whether they made economic sense or not."

Up to this point we have focused on the private interests of specific institutions and communities. But the development of hydraulic projects in the American West over the course of the nineteenth and twentieth centuries has to be understood in far broader terms than the mere summation of the activities of individual actors. As Öjendal writes, in his review of work by Wittfogel, Toynbee and Caponera, "water is a key element... in building a society". There is no lack of material to substantiate that higher level vision in the development of the American West. Nor is it necessary to defend or embrace such philosophies. The point is to recognize that they existed and that they were powerful ideas in shaping the nation. Below we set out some examples of these ideologies.

President Jefferson believed that a nation of small farmers is a nation with a purer soul. In the 1880s greening the desert became a kind of Christian ideal. Also in the 1880s, Senator Thomas Hart Benton fathered the concept of Manifest Destiny, a destiny that could only be realized if floods were controlled, rivers tamed for navigation, cities provided with water for industry and households and water supplied to farms. In the twentieth century President Theodore Roosevelt supported the Reclamation programme because he saw it as an agrarian path to industrial strength.

The development of California and a great city where Los Angeles stood was seen as giving the USA the capacity to protect its western flank against the ascendant Orient-Japan in particular. President Franklin Delano Roosevelt during World War II feared the return of the inter-war depression and saw water resource projects as a vast job-creating engine just as he had used public works of many kinds in the 1930s New Deal. In 1935 he had ordered the Bureau of Reclamation to take over the CVP to provide 100,000 jobs for migrants displaced by the Dust Bowl. In Washington State he had backed the Grand Coulee dam as a source of public-sector hydropower and as an employment source for tens of thousands of construction workers and, thereafter, agricultural settlers.

And so we return to social cost-benefit analysis. *Cadillac Desert provides no evidence whatsoever that this evaluative technique, first developed in the United States itself during the 1930s, played any significant role in water resource project choice.* In practice, dam development was driven by the twin motors of the particularist motivations of individuals and institutions that gained from dam construction and of expansionist ideologies. As Carruthers and Clark write:

> It is not too cynical to note that cost-benefit analysis, as presently practised, had its origins in the water sector in the 1930s in the United States where various government agencies were anxious to promote worthwhile public projects which did not appear sufficiently attractive with formal financial tests. Repeatedly we learn of projects approved in spite of the weakness of their economics. Irrigation ideas were "...pursued with near fanaticism, until the most gigantic dams were being built on the most minuscule foundations of economic rationality and need". One of the projects on President Jimmy Carter's hit list of projects that he wished to emasculate returned only 5 cents in each dollar invested. The economics of the Colorado River Storage Project were dismal: high-altitude desert farmers were being subsidized to grow the same crops that Illinois farmers were being paid *not* to grow.

The committees of Congress asserted or implied that billion-dollar dams of Central Arizona and Tennessee-Tombigbee made little sense. Weak projects were slipped onto omnibus authorization bills. During the five-term Roosevelt-Truman presidential years, several river-basin bills authorized dozens of dams and irrigation projects at a single stroke. Economics mattered little, if at all; if the irrigation ventures slid into an ocean of debt, the huge hydroelectric dams authorized within the same river basin could generate the necessary revenues to bail them out.

Mike Straus, the Bureau's Commissioner, is quoted as saying to a group of employees in Billings, Montana, who were responsible for the upper Missouri Basin: "I don't give a damn whether a project is feasible or not. I'm getting the money out of Congress and you'd damn well better spend it. And you'd better be here early tomorrow morning ready to spend it, or you may find someone else at your desk."

This is not to suggest that cost-benefit analyses were not prepared-far from it. But too often they were ignored or doctored. Engineers would grotesquely underestimate costs. Once a project was approved, a design change would result in much higher outlays, as in the case of the Grand Coulee. A senior employee is on record as having to 'jerk around the benefit-cost numbers' to make "trash" projects look good. A bad project would be compared with a worse one so that it could be argued that the former was the least-cost alternative.

The forced resettlement of an Indian reservation was counted as a benefit because the Indians would get to live in a "nicer" town. In the case of the Teton dam it is argued that the Bureau overstated the flood control, wildlife and fishing benefits, exaggerated the number of new acres opened to irrigation and promised supplemental water of an average of 5 inches (127 mm) to farmers already using an average of 132 inches (3,353 mm).

The dam was approved. The dam was built. At this point that great evaluator in the sky took a hand. On the same day as it was filled, the dam collapsed and was swept away, killing in the deluge eleven people, damaging or destroying 4,000 homes and bringing the loss of 350 businesses.

THE NEED FOR MULTI-CRITERIA ANALYSIS

Social cost-benefit analysis is a useful tool that should always be deployed in the evaluation of capital projects in the irrigation and drainage sector that are financed by government. However, the work of Carruthers and Clark (1981), Marc Reisner (1993) and the WCD (2000), to cite just three examples, convinces that SCBA should never be the only form of evaluating large projects.

This is because, in spite of its name, SCBA is too narrowly focused on *economic* outcomes. All projects have costs and benefits that are economic, social and environmental. SCBA, if properly carried out, has the capacity to deal with the *economic* dimension in respect of project-triggered changes in GDP. With the second category, *social* effects, we refer to all those impacts on individuals, families and institutions that are not already dealt with under the economic heading.

The social category is extremely broad, ranging from outcomes such as the displacement of an Indian tribe from its homelands through to the agricultural and industrial development of an entire region. The third type of impact is *environmental,* where a capital project brings about the destruction or degradation of local or regional natural habitats. The big question is how best to make decisions on such projects, given the variety and heterogeneity of their impacts? We suggest seven golden rules of procedure:

- The evaluation of a defined capital project should always be based jointly on its economic, social and environmental effects in comparison with the noproject alternative.
- The fantasy of economists that the social and environmental impacts can be meaningfully evaluated in dollar terms should be dismissed. This challenge to environmental economics is developed fully in *The Economics of Water Resources.*
- The appraisal should be honest.
- The evaluation should be based on values as well as on qualitative and quantitative information.
- At the country or region level, the decision to proceed (or not)

with a project should be taken by a group, balanced in its interests and skills, that understands and represents the economic, social and environmental categories of benefit and cost. Engineers should be represented on that group since their expertise in the *how* of the project and technically feasible alternatives is invaluable.

- The group should indicate in its conclusion the relative weight given to each of the three categories in arriving at its decision as well as the principal means by which each category of impacts is evaluated.
- The information on which a decision is based should be comprehensive, intelligible and readily available to all citizens.

Chapter 10

Water Allocation at the Regional Scale

A THOUGHT EXPERIMENT

Imagine for a moment that you have the power, given by God or science, to observe over a full year in any region of the world each and every molecule of water appropriated by human society ("input" molecules). Imagine too that in that region in that year you can observe how each such molecule is used within human society or lost to it ("output" molecules). An extraordinary gift! You would then be able to deploy a water resource planning tool that we call the region's "hydrosocial balance".

The hydrosocial balance is applicable to *any* defined geographic space; the term "region" is used advisedly because of its inherent ambiguity. A region could be a continent, a country, a province, a catchment, an irrigation district, a city, a village, a large sports facility, or the site of a manufacturing firm such as a sugar mill.

The hydrosocial balance is also always applied to a defined time-period. For convenience of exposition it is assumed here that the time-period is the year 2002. Alternatively it could be for the five years 1998-2002, for example, or for the month of August averaged over the 10 years 1993-2002. The hydrosocial balance for a past or present time-period is referred to as a *baseline balance;* one for a future time-period is a *scenario balance.*

The generic form of the hydrosocial balance. *Let us first consider the supply-side categories.* Rainwater collection, ground water and surface-water abstraction, desalination plant output and the import of water from other regions are all input molecule flows requiring no further commentary. The annual supply they provide can be measured in megalitres per day or millions of cubic metres per year or litres per head per day (lhd) of the resident population.

Internal reuse occurs when a household or a factory or any other organization reuses its own wastewater. We have already seen that the pumping of agricultural drainage for irrigation purposes is an example of such a supply form. The water volume of internal reuse is set equal to each cubic metre of freshwater supplied to the user multiplied by the number of times it is reused. External reuse occurs when the wastewater of one organization or group of households is reused by a separate body, as in the reuse of treated sewage by agriculture.

Supply leakage and evaporation, as well as the export of water to other regions, are also examples of supply-side flows. But these are output flows and for that reason are entered as *negative* items in the supply.

Next let us consider the use categories, all of them output molecule flows. The entries are fairly conventional in their breakdown and require little comment. Only in the case of "instream applications" is it worth explaining that these refer (for example) to surface-water abstraction that is thereafter returned to the hydrological resource as part of some conservation plan, such as the management of a wetland. Note too that use is inclusive of beneficial use, reuse volumes and leakage, evaporation and wastage on user properties.

The thought experiment that gives us the power to track the movement of each and every molecule in the system now comes into play. Every input molecule that enters the region's hydrosocial balance in a given year, whatever the supply source, will then reappear as an output molecule in terms of how it is used or how it is lost to supply. We have in place an accounting system for outstream water in which every molecule is counted twice, once as input and once as output, just like the double-entry book-keeping of the accountancy profession.

Up to this point, the balancing of the water accounts ledger has derived from the notion that each and every molecule of water recorded as an input to the regional system is then recorded as an output from the system. The question then arises: does the existence of water storage infrastructures destroy the accounting balance? Our understanding of this issue will be strengthened if we imagine the region's storage capacity in the year 2002 as being composed of just three reservoirs: one black, one gold and one green.

The black reservoir is dedicated to the storage of water abstracted in 2002 that in the same year is distributed in its entirety to users or lost to leakage, evaporation and export. Clearly, storage of these pass-through molecules in the black reservoir does not change the systemic balance for 2002.

The gold reservoir contains stored water from time-periods prior to 2002. These molecules are, so to speak, a gift from the past to the present. In 2002 some of the stored water is lost to supply leakage and evaporation, some may be delivered as water exports to another region and some is distributed to users. The fall in the quantity of water stored in the gold reservoir during 2002 is expressed in megalitres per day and is deemed to be an input molecule flow. Once again, the identity of total net supply and total use is maintained. The green reservoir is dedicated to the receipt of water abstracted during 2002 that will be stored for distribution from 2003 onwards, a gift from the present to the future. The volume stored in the green reservoir during 2002 is expressed in megalitres per day and is deemed to be an output molecule flow. This flow precisely matches the abstraction flow pumped to the reservoir and, for the third time, the mathematical identity holds.

In practice, of course, each reservoir in a real regional system combines the functions of all three reservoirs described above. What we observe is only the net outcome of the component processes, i.e., either no change in 2002 in the volume of stored water, or a fall or a rise. Thus, with respect to the value of "K" no change in the total volume of stored water gives a value of zero, a fall is expressed at its daily rate and is recorded as an input molecule flow and a rise in storage is expressed at its daily rate and is recorded as an output molecule, negatively valued flow. The analysis here of reservoir storage applies with equal force to the social processes of aquifer recharge and recovery.

It follows that in any base year for which a region's hydrosocial balance is calculated, even without the advantage of the gift to track every molecule, total net supply is always equal to total use, *provided* that the regional water resource planners' data are comprehensive and accurate.

WATER RESOURCE PLANNING AT THE REGIONAL SCALE

The hydrosocial balance may be a baseline balance for a past or present year or it may be a scenario balance for a future year. When the baseline balance and the scenario balance have the same structure, we can subtract the entries of the former from the entries of the latter to give a *hydrosocial change balance.*

Let us outline the potential value of such balances and suggest why they may become routine tools of water resource planning in a wide range of countries over the next decade. *The baseline balance* can be prepared for any geographic area and provides a comprehensive, synoptic account of the scale and composition of the supply sources of water and its uses in that region. The approach is pragmatic. One can foresee an evolution in statements for a specific region from the simplest set of water accounts to more complex ones, as policy interests drive the expansion of the categories deployed and the accuracy of their measurement. The work process to make the baseline calculations will stimulate new directions for research and will deepen planners' understanding of the region's hydrosocial cycle within its hydrological context. The four-column approach gives equal weight to both supply-side and demand-side information and describes them in the language of a unified conceptual framework. The reuse entries will strengthen our understanding of two of the cyclical components of water resource planning-make them more visible. The fundamental identity of total net supply and total use provides a consistency check on individual calculations, akin to the trial balance procedure of company accountants.

The change balance, the planning tool *par excellence,* is applicable to short-, medium- and long-term strategy–to compare of typical years past and future–and it clarifies the quantitative choices that a regional government or catchment agency will be forced to envisage, for example, in a future year of severe drought or as a result of climatic change. The change balances can provide comprehensive, synoptic, transparent alternative scenarios within a region in planning infrastructural investment, capital financing and demand management. It is in this way that it handles the uncertainty that, as Peter Rogers writes, is a major issue in water resources planning, as well as side-stepping the inadequacy of project-by-project planning. The change balance does not evaluate a scenario, this is the work of environmental scientists, economists and others, but it does impose a consistency check on each case, because of the fundamental identity of total change in net supply with total change in use.

Now let us suppose that in a specific region, for a 5- or 10-year planning period, water resource managers developing alternative hydrosocial change balances become aware that there is a serious impending conflict. The potential conflict is between irrigation as a user of water on the one side and the non-farming users on the other, as well as the instream requirements for lake and river flows-nature conservation, fishing, navigation and recreation. An era of allocation stress is fast approaching.

Group I consists of countries that face physical water scarcity in 2025. This means that, even with highest feasible efficiency and productivity of water use, these countries do not have sufficient water resources to meet their agricultural, domestic, industrial and environmental needs in 2025. Indeed, many of these countries cannot even meet their present needs. This category includes countries in the Middle East, South Africa and the drier regions of western and southern India and northern China and contains 33% of the total population.

The only options available for these countries is to invest in expensive desalinisation plants and/or reduce the amount of water used in agriculture, transfer it to the other sectors and import more food. The degree to which they can increase the productivity of water depends very much on the possibilities they have to change their irrigated staple food production to high value crops or to other more productive uses. This requires access to markets for farmers and countries for these higher value products to enable payment for the food crop to be imported or bought locally. An alternative is the import of water into the region through inter-basin transfers as being considered in India and northern China.

Conflict resolution over allocation stress may be possible through substantial additional investment in supply-side expansion, it may be feasible through demand management in the water-using sectors and it may be achieved through reuse and recycling strategies including aquifer recharge. But, to press the point, suppose the allocation of water to the irrigation sector is under review, with a view to reducing its absolute volume and its percentage of total use, for just such a circumstance.

Irrigation Water Efficiencies

Relative efficiency is an agronomic concept and a tricky one at that because it is the ratio of two ratios. It can be used to show that, for a wide variety of crops, proportionate increase in water supply is matched by proportionate increase in crop tonnage but that this response fades at higher levels of water supply. By implication there may be little or no agronomic or economic justification in seeking maximum plant yields.

Requirements efficiency is a distributional management concept. Assuming that a crop's net irrigation requirements are in fact met, it indicates the base supply actually used to resource those requirements. For example, at a value of 1.0 the whole of the base supply is taken up by crop evapotranspiration; at a value of 0.5 the base supply is twice the net irrigation requirements.

Tonnage efficiency is a planning concept of limited interest, a ratio of crop output to the base supply of irrigation water. It is a close cousin of the farmer's concept of yield-output tonnage per hectare. Tonnage efficiency can be used in predicting the water resource requirements of the agricultural sector in a future year. In itself it is of little economic interest since it reflects neither output value nor input cost.

Marginal tonnage efficiency shows the relation between output increase in tonnes and an addition to the base irrigation supply volume. Field work often suggests it has a zero value. When the value is known either to farmers or to agricultural extension workers, it can be useful in moderating the demand for irrigation water.

Marginal turnover efficiency is a concept of orthodox economic theory, the relation between marginal revenue and marginal cost in respect of irrigation water supply. But as a determinant of farmers' behaviour it is likely to be of importance only where they bear the cost of additional water supplies, where they have control over the volume of water they use and where the supply of water is a significant component of prime costs. *Gross margin efficiency* was introduced in the case study of the Curu Valley, where Karin Kemper called it net gain or net return and used it to show the average monetary return in terms of the gross margin per cubic metre of irrigation water. It is calculated for different crops but in the Curu Valley did not explain crop composition. *Turnover efficiency* appeared in the Keveral Farm case study and is equal to the gross sales income of irrigated produce per cubic metre of irrigation water. It can be estimated at the farm, scheme and irrigation district levels.

Field supply efficiency, as in the case of E_2 above, this measure contains no variable representing agricultural output. It gives the sum of the average prime cost and the average overhead cost for delivering 1 m^3 of water to the farmer's field. Its uses are in understanding the efficiency of water supply infrastructures that generate this supply and in designing water charges to be levied on farmers. *Marginal field supply efficiency*, as with E_8, this is equal to average total cost of field supply. But in this case it refers to any specific new investment in the irrigation infrastructures of a region and the additional volume of water that is thereby delivered.

Net benefit-investment ratio is most favoured economic measure for the evaluation of irrigation and farmland drainage projects. If calculated in the correct way, it is of great value to infrastructural planning. Most of the ten concepts above are useful in particular circumstances. But what we are searching for here is a variable that enables us to compare *water efficiency in different sectors of the economy*. Presuppose that the allocation of irrigation water is under review at the regional level with a view to its transfer to other economic sectors. We therefore need an efficiency measure that enables us to judge whether such a reallocation is justified on the basis of relative sectoral efficiency in water use. In fact an eleventh measure of efficiency is necessary:

$$E_{11} = \frac{V_i}{S_{bi}}$$

where V_i is the value added in a defined economic sector i and S_{bi} is the base supply of water to that sector. Value added is the economist's measure of the flow of output. In any enterprise (including farms) the annual flow of value added is the difference between the total annual receipts from sales and the total annual cost of bought-in materials upon which the enterprise's labour and capital work. At the sectoral level it is the aggregate value added of the enterprises that make up that sector. We shall refer to E_{11} as "sector efficiency".

Note that when distinguished authors refer to "crop per drop" without defining the term, they probably mean tonnage efficiency or marginal tonnage efficiency. Similarly, when they refer to "value per drop" they may mean gross margin efficiency or turnover efficiency or sector efficiency. It may even be the case that they do not know what they mean.

On both *a priori* grounds and those of practical experience we would expect that the irrigated farming sector would exhibit low sector efficiency. This is because irrigated agriculture is inherently an intensive user of water. Low sector efficiency will usually be associated with a low ratio of employment to water supply. The conclusion is that in a regional review of water reallocation between sectors purely on efficiency grounds, irrigated agriculture is likely to rank last. On this criterion irrigated agriculture would have its allocation constrained or reduced.

This flow derives from leaching applications, escapes, seepage, deep percolation and surface run-off, wherever these flows are not lost to evaporation. To put it another way, the consumption of irrigation water in crop evapotranspiration and in non-beneficial transpiration and evaporation never exhausts the base supply. That being the case, a proportion of the volume of irrigation water measured in the sector efficiency variable, *S*bi, becomes available for use by other sectors on its return to the hydrological resource as drainage.

As far as it goes this point is valid. In the appraisal of irrigation projects such as canal lining or flood irrigation it can be of great importance. However, in the comparison of sectoral efficiencies it is a bad defence of agriculture. This is because the sectors with which agriculture is competing for water *also* recycle their wastewater to the hydrological resource shows. Moreover the proportion they recycle is greater than that of irrigated agriculture. The recycling counter-argument does not rebut the case against irrigation, but reinforces it.

With the assistance of colleagues in the International Commission on Irrigation and Drainage (ICID) we have pursued this question of the sectoral consumption of global outstream water. The key datum is that irrigated agriculture consumed in 1995 an estimated 93 per cent of the world's abstraction of water for outstream purposes–1,750 billion cubic metres.

SUSTAINABLE RURAL LIVELIHOODS

The assumption is made that a regional water resource management agency is reviewing the existing allocation of water to irrigated agriculture with a view to reducing its absolute volume and its percentage of total use. Since the mid-1990s, university researchers and aid policy personnel have devoted much energy to a new approach in the understanding of rural change under the rubric of "sustainable rural livelihoods". This mode of analysis begins with an emphasis on *people* and how they can work together to build on their own strengths with the objective of eradicating poverty. The focus is rural yet rural/urban linkages are emphasized and related to livelihood diversification over time. There is no assumption that because rural people are farmers they will stick with this in perpetuity to the exclusion of other strategies. As Diana Carney writes:

> The sustainable rural livelihoods agenda ... embraces the fact that rural people's livelihoods are very diversified, that they are intimately connected (in both positive and negative ways) with the livelihoods of urban dwellers and that the prospects for advancement and for increased robustness of rural people may not always lie in the sectors with which they have been traditionally associated.

Table:1. Global Water Withdrawals and Consumption by Economic Sector in 1995

	Withdrawals		*Consumption*	
	km³/year	%	*km³/year*	%
Agriculture	2,500	70	1,750	93
Municipal	340	9	50	3
Industrial	750	21	80	4
Total	3,590	100	1,880	100

At the heart of the sustainable rural livelihoods framework lies the concept of five types of capital asset upon which individuals draw to build their livelihoods:

- Natural capital refers to the natural resource stocks from which resource flows useful for livelihoods are derived. In the context of this book, natural capital is the hydrological resource.
- Social capital refers to the social resources (networks, membership of groups, relationships of trust, access to wider institutions of society) upon which people draw in pursuit of livelihoods.
- Human capital refers to the skills, knowledge, ability to work and good health vital to individuals' capacity to pursue different livelihood strategies.
- Physical capital is the basic infrastructure in transport, shelter, energy, communications and water, as well as the sectoral production equipment and stocks that enable people to pursue their livelihoods.
- Financial capital refers to the financial resources that provide people with different livelihoods.

These capital assets can be represented diagrammatically as a pentagon where access by a household or social group to each of the five capital asset categories can be plotted, whether access is in terms of private ownership recognized in law or through customary use rights. Carney suggests such a pentagon challenges those that use it to think holistically about asset combinations as the basis of livelihoods. Intuitively, she writes, there is a close correlation between people's overall asset status and their robustness in the face of adverse external trends and shocks to a specific way of life. Building up assets is a core component of empowerment.

The sustainable rural livelihoods framework also draws upon the concepts of organizations (from layers of government through to the private sector and its markets) as well as processes (politics, laws, rules of the game, incentives). The parallel with the institutional character of political economy is clear here.

Organizations and processes are critical in determining who gains access to which type of capital asset and the effective value of that asset. In conjunction with people's asset status, they also shape the livelihood strategies that are open and attractive to individuals, households and social groups. Markets, in particular, profoundly influence the convertibility of one asset into another. Such convertibility can increase people's options, help them improve their livelihoods and reduce their vulnerability to negative trends and shocks. But counter-examples also exist of the effects of market power. Soussan writes "the development of local water markets can prove to be highly inegalitarian and subject to capture by elites. Successful livelihood strategies raise group incomes, increase their well-being, improve food security and, in an environmental sense, make use of the natural resource base sustainable".

In brief, the analytic framework of the sustainable rural livelihoods approach to change in rural areas can be said to be made up of three principal blocks of ideas. *First,* in their daily lives individuals draw on natural, social, human, physical and financial capital asset combinations. *Second,* society's organizations and processes powerfully influence people's access to capital assets and these assets' values. *Third,* in shaping their livelihood strategies to build a life and endure in the face of adverse trends and shocks, people's real options are fundamentally shaped by their asset status and the institutional framework.

In the light of this approach, what can we say about a regional water resource policy that reduces significantly the volume of irrigation water supplied to the farming sector? Clearly here we posit an adverse shock and a continuing negative trend for rural livelihoods. Farmers' access to the natural capital asset of the hydrological resource is cut back. At the same time, the values of assets such as farmland diminish. Farmer networks may be weakened as some suffer whilst others escape unscathed. The human capital embedded in the daily routines of agricultural practice is devalued. Physical capital becomes obsolescent overnight. Financial capital is eroded as income from the sale of produce disappears.

The rapid decline in farming-sector households' and businesses' assets undermines their feasible livelihood strategies. They may continue in farming but at lower output levels and on reduced incomes. They may become labourers for others or seek work in rural construction and other such industries. Some might resort to crime and banditry. Others, perhaps the majority of those who have been hit hard by this change in the social structure, may drift into the towns and cities in search of an urban livelihood. This will take place more rapidly when families already benefit from existing rural-urban linkages.

The implications of this scenario for regional government are clear. Where its forward planning, using the hydrosocial change balance, indicates serious conflict over water allocation, government action must include one or all of the following measures:

- Review the possibility of expanding the water supply infrastructure, at the same time as taking fully into account the financial and environmental consequences of such projects.
- Push through management structures and practices that reduce basin supply losses and raise water efficiency in use in all sectors–agriculture, industry and households.
- In particular, pursue action raising irrigated farming's sector efficiency (E_{11}).
- Ponder long and hard about imposing adverse shocks on the irrigation sector, because of the devastation it can wreak on sustainable rural livelihoods, accompanied no doubt by widespread, violent unrest.
- In the case where, in spite of all the counter-actions and counter-arguments, it is judged that allocation *should* be reduced, regional government must develop its settlement, industrial and training infrastructure. Thereby new economic opportunities are opened up in villages, towns and cities to those who no longer engage in irrigated farming.

NATIONAL FOOD SECURITY

The third and last debate we shall consider with respect to the reallocation of irrigation water to urban use is that of national food security. But before launching into this review of the case for and against irrigation, it is necessary to clarify how the term "national food security". Consider the situation in two countries, A and B:

- In country A the agricultural sector has production capacity in a year of average harvests to meet in full the food needs of the resident population.
- In country B the agricultural sector can meet only a proportion of the resident population's food requirements but the sector's exports in value terms are sufficient to permit country B to import all its residual food needs at the ruling level of world prices. 'Residual food requirements' are total food needs less those met out of domestic production.

Any country in which either 1 or 2 above holds true will be regarded here as enjoying national food security. Note that situation 1 (but not situation 2) is widely referred to as food self-sufficiency. The existence or absence of food self-sufficiency can be demonstrated in detail by developing a food balance sheet for the region, as Foster and Leathers show. The food balance sheet is the solid cousin of the waterish hydrosocial balance. Furthermore, Foster and Leathers show that food security is also deployed at the level of groups of households.

Suppose we are in a country that is just on the margin of national food security, just in balance in terms of either criterion 1 or 2 above. Here the National Water Resources Agency proposes a 10-year plan, the effect of which will be massively to shift water out of irrigation into urban use, where E_{11}, sector efficiency, is strikingly higher. Let us consider in turn the case *against* and the case *for* such reallocation.

The case *against* reallocation (and therefore the case/or irrigation) is that such a policy will drive the country into a situation where its agricultural sector is not capable of meeting the nation's food requirements, directly or indirectly.The argument continues: the fundamental duty of the State is to ensure that a people, a nation, can survive. The abandonment of national food security is a breach of that duty. The case for reallocation completely accepts the argument that the State must ensure the capacity of a nation to survive. It is also admitted that in the new situation after reallocation the country neither will be able in an average year to feed itself, nor will agricultural exports be sufficient to cover residual food requirements.

But the counter-attack comes quickly. Yes, residual food needs must be met, but the foreign exchange needed to make that possible by no means has to be met from agricultural sector exports. US dollars, pounds sterling, euros, or what you will, can be earned by the exports of the mining, manufacturing and tourism industries to which the reallocated irrigation water is being diverted. Metaphorically, the water lost to domestic agriculture as a result of the National Water Resources Agency's 10-year plan will be substituted by the virtual water of the countries exporting the products that satisfy our example country's swelling residual food requirements. This is Tony Allan's (2001) approach to the issue. After all, if a capital city such as London can flourish and grow so spectacularly for 2,000 years to become for those that love it the greatest city in the world, *always in a situation of complete regional food insecurity*, why should not the same "insecurity" be consistent with the long-term flourishing of a country?

When set out in these general terms, the case for reallocation seems very strong. However, national policies are never made and never should be made on general grounds. The contingent and historically specific circumstances must always be appraised alongside the general rule.

Consider again our example country on the cusp of national food security. It should certainly review whether reallocation would be accompanied by a *loss* of export earnings, as well as gains, where the country either exports or could export agricultural produce and manufactures based on such products.

Alongside its 10-year water resources strategy, the country should review the long-term prospects for its dollar exports of raw materials, manufactures and tourism services. The country should also review likely trends in the prices of imports, particularly food. If we have a case where reallocation will lead to losses of foreign exchange earnings, if non-food export prospects are not favourable and if there is a long-term upward trend forecast in the price on the international market of food imports such as wheat, maize and rice, then the good sense of water reallocation to urban uses may be open to strong challenge.

Particularly in arid and semi-arid countries with high rates of growth of population, globalization can produce a situation where a country is chronically and deeply dependent on food imports, where the growth in export earnings is weak and where its leaders seem to have no alternative other than to turn to the USA, Canada and the European Union for food aid. Such gifts always carry political debts that may prevent a country from pursuing policies that do not accord with the one-size-fits-all dreams of some theoretical, macro–economic scribbler in a distant tower. Van Hofwegen and Svendsen also point to the importance of effective food distribution systems in the debate.

Where imported food is available, distribution is often problematic because infrastructure for transportation and administration is inadequate. Moreover, people's access to food depends on their income. Currently more than 1.3 billion people are absolutely poor, with incomes of a dollar a day or less per person, while another 2 billion people are only marginally better off. Income growth rates have varied considerably between regions in recent years, with Sub-Saharan Africa, West Asia and North Africa struggling with negative growth rates. In such regions people's own capability to feed themselves is of utmost importance. Rural development is essential and security, basic infrastructure, improved technology and better land and water management practices are important conditions to achieve it.

To end these analytic sections, it remains to be stated what are the means to carry through allocation away from the irrigated farming sector, if this is the decision of the regional or national Water Resources Agency. Reallocation can be achieved, when such a policy is politically feasible by:

- Using abstraction licensing to reduce abstraction for irrigation purposes.
- Shifting the water charging system in the irrigation sector from a zero charge or revenue-only fee to one of volumetric pricing or indirect fees.
- Developing a regime for tradeable abstraction rights operating between the irrigation sector and the urban water-using sectors.
- Allowing urban encroachment into the irrigated farming areas, with compensation paid to farmers for the loss of their land.

WATER ALLOCATION ON THE PALESTINIAN WEST BANK

The West Bank has an area of 5,800 km^2. In 1997 its population totalled some 2,230,000 persons, inclusive of about 170,000 Israelis living in 150 settlements as well as another 180,000 Israeli settlers living in East Jerusalem. The average household size in the territory is 6.1 persons and the population density is some 385 persons/km^2. The West Bank has hot and dry summers but cool and wet winters. Two currencies are in use- the Jordanian dinar and the New Israeli shekel: at the time of writing, in March 2001, the rates of exchange with the US dollar were, respectively, 0.71 and 3.45. Shops also accept US dollars.

The West Bank is an occupied territory and Israel is the military occupier. As a result the political system there is *sui generis*. The signature of the Declaration of Principles in September 1993 by the Palestine Liberation Organization and by Israel led to the establishment in 1994 of the Palestinian Authority, composed of the West Bank and Gaza. This later became the Palestinian National Authority (PNA). In 1996 the Palestinians held elections that made Yasser Arafat President and created a Legislative Council with eighty-eight seats.

But the interim accord between the PNA and Israel gives the West Bank only limited self-rule. The most critical issues in the final status negotiations are the borders of an independent Palestinian state, the status of Jerusalem, the future of the Jewish settlements in Gaza and the West

Bank, the right of 3.6 million Palestinian refugees to return to their homeland and access to water resources. In respect of the last of these issues, the most important PNA institutions are the Palestinian Water Authority (PWA) and the Ministry of Agriculture.

Access to water resources is composed of capture of groundwater springs, penetration of the mountain aquifer's western, north-eastern and eastern basins by wells and boreholes and abstraction from the River Jordan. But such an assumption would be mistaken. The Upper Jordan is fed by the Dan spring, the Hasbani River, the Banias River and other headstreams prior to its entry to Lake Tiberias. South of the lake the river is joined by its major tributary, the Yarmuk. Further downstream still, winter flood waters flow along the wadis, primarily from the east bank and the most important of these tributaries is the Zerqa. In the absence of human abstraction, the river's average annual flow is estimated at most as 1,850 mcm/ year at the Allenby Bridge on the Lower Jordan, whereas other estimates are as low as 1,100 mcm/year. Tiny by world standards, the Jordan basin is 18,300 km^2 in extent, 3 per cent of which lies within the State of Israel and a similarly small percentage within the area of the West Bank.

Schwarz reported that of the total 610 mcm flow from the Upper Jordan to Lake Tiberias, the Israelis were appropriating 570 mcm, either directly from the Lake or from the flows sourcing it. As a result there is no freshwater flow out of the Lake into the Lower Jordan, only saline waters issuing from the Lake's springs. In addition much of the flow of the Yarmuk is captured by Israeli abstraction and by the King Abdullah Canal in Jordan. Like the Yarmuk, the Zerqa also is used for Jordanian irrigation and other demands.

Along the West Bank, therefore, the River Jordan-of biblical fame-is nothing more than a streamlet. Moreover, Palestinian access to the west bank of the River Jordan is prohibited by Israel, which denominates land along the Lower Jordan as a 'closed military area'. The river meanders, so the military order forbids Palestinian farmers from entering a zone, which may mean a distance from as much as 5 km but at least 1 km from the Jordan's bank. Israeli farmers can cultivate right up to the river-bank. It is of some interest that the much-respected Johnston Plan of 1955 implied the West Bank's share of the waters of the Jordan should be 140-150 mcm/ year. The Palestinian Ministry of Planning (1998) proposes that some 560 mcm should go to Jordan and 215 mcm to the West Bank.

In respect of groundwater, the Western basin flows towards the west, the North-Eastern flows northwards under the Vale of Esdraelon and the eastern flows towards the River Jordan. These basins (principally the first two) extend into Israel under the "Green Line", i.e., the line originally defined by the 1949 Armistice Agreement between the states of Jordan and Israel. The average annual recharge of the three basins through precipitation is some 650 mcm, 80-90 per cent of which falls on the Palestinian side of the Green Line. The accuracy of the recharge figures is uncertain.

The June 1967 war between Israel and the Arab countries brought the West Bank–and therefore its basins–for the first time within the control of the Israeli state. At that time Palestinian abstraction from the Western and North-Eastern basins was about 30 mcm/year and a further 50-70 mcm/year was taken from the Eastern basin. Israeli abstraction within Israel was about 300 mcm/year. After June 1967 the Israelis introduced to the West Bank a licensing system for well-drilling and well-expansion. Shapland writes: "In practice, licences have only been granted in a few cases, all of them for domestic water; no licence has been granted for the provision of water for agriculture." These constraints were maintained despite the continued growth of the Palestinian population. Meanwhile Israeli abstraction within Israel continued to grow, to about 560 mcm in 1999. However, the suppression of Palestinian demand was not matched by parallel constraints on Israelis erecting new settlements in Gaza and the West Bank. In the latter case they derive their water from wells drilled into one or other of the three mountain basins. More than half comes from the deeplying Lower Cenomanian aquifer.

These brief paragraphs serve to provide the hydrological, hydrogeological and political context that shapes the West Bank's hydrosocial balance. The principal source is a set of data brought together by the Palestinian Water Authority and the University of Newcastle upon Tyne during the early stages of a joint project with the British Geological Survey on the sustainable management of the Gaza and West Bank aquifers, funded by the UK's Department for International Development. At the time of writing the material is in draft. In the case of *rainwater collection,* this takes place through the use of cisterns and small, shallow ponds. The flow has been estimated as 7 mcm.

Palestinian groundwater abstraction from boreholes in 1999 equalled 61 mcm. Note that all these figures are rounded to the nearest million cubic metres but that even this overstates the information's accuracy. The datum is entered on the left-hand side of the cell to indicate that it will be entered into a subtotal. Palestinian groundwater abstraction from springs is hardly different from the borehole subtotal. Total Palestinian groundwater abstraction is given both in millions of cubic metres and in litres per head per day. The 1999 Palestinian population of the West Bank is assumed to be 1.9 million. The litres per head per day data are rounded to the nearest 10 in endeavouring to avoid an impressive but spurious accuracy.

Jewish settler groundwater abstraction in 1999 from boreholes was 42 mcm and from springs was 100 mcm. But this last datum is at best an educated guess: it refers entirely to the capture of brackish water from springs in the Eastern Basin to use in the irrigation of crops in the Jordan Valley. The datum for total Israeli abstraction of groundwater in litres per head per day uses a population estimate of 350,000 settlers in the West Bank and East Jerusalem.

The disparity between Palestinian groundwater abstraction at 180 lhd and settler abstraction at 1,110 lhd is prodigious. As we have already seen, it is explained by the fact that, since the 1967 Israeli military occupation of the West Bank, Palestine abstraction has been severely constrained whereas appropriation by Jewish settlers is "apparently unrestrained" *Surface-water abstraction* from the River Jordan does not take place. The only other surface-water source for abstraction is the wadis, such as the Wadi el Farah But I know of no statistic on such abstraction.

The desalination of salt or brackish waters does not take place anywhere on the West Bank, as far as we are aware.

With reference to the *import of water from Israel,* there are pipelines running between Palestine and Israel in the area of Hebron, Bethlehem, Ramallah and to the west of Ramallah. We know that there are large numbers of wells to the immediate west of the border between the West Bank and Israel, so it seems likely that Israel does indeed export pumped groundwater from the Western basin back to the West Bank whence the water first originated as precipitation. The Jerusalem Water Undertaking purchases 25 mcm/year from the Israelis and Shapland suggests that the Jewish settlements receive some of their water piped from the Israeli water system, Mekorot. But West Bank import of water from Israel is a total not known to me. The water from Israel and from Israeli-controlled wells on the West Bank does not always come from a single source; it is often mixed prior to distribution.

In respect of the *internal reuse ofwastewater,* we know that the West Bank's urban areas repeatedly experience water shortages. So it is likely that urban households and industry, as well as rural households, reuse their water internally. For example, small firms may reuse waste water for some cleaning purposes and families are likely to reuse wastewater for a kitchen garden. The volume of such use is probably high–but not known. Internal reuse in agriculture is probably negligible because the drainage flows in the irrigated sector would be too sparse to justify pumping them back to the cultivated area.

External reuse of wastewater on the West Bank refers to the plans to collect, treat and distribute to the agricultural sector the urban wastewater of Al Biereh/ Ramallah and Salfit. The process had not yet begun in 1999.

The pipelines referred to with respect to the import of water from Israel could be a means of *exporting water to Israel*. David Scarpa of Bethlehem University suggests that a considerable amount of water is exported in this way (personal communication). But the total is unknown to us.

It is now possible to give a crude estimate of the *supply and distribution losses* between points of abstraction and points of use. In the case of groundwater abstraction for agriculture, it is probable that seepage and evaporation losses prior to water reaching farmers' fields are small, because of the short distances involved. Such losses are here assumed to be equal to 10 per cent of gross supply to that sector. In the case of supplies to households and urban uses, losses are likely to be much higher, of the order of 35 per cent. We can estimate the gross groundwater supply split between agricultural and non-agricultural use from the Palestinian Water Authority data and therefore derive total losses. These are 36 mcm/year (44 lhd).

It is worth noting here that the Deutsche Gesellschaft für Technische Zusammenarbeit (GTZ) in 1995 estimated that 70 mcm is lost from flooded surfaces including the four major permanently flowing wadis of the West Bank. But these, of course, are hydrological flows, not the hydrosocial losses.

The final category of supply *is fall or rise in volume of stored water*. We do not know the number or volumetric capacity of the West Bank's surfacewater reservoirs. On both counts the figures are likely to be small. Moreover the hydrosocial balance requires only the *difference* in the volume of abstracted water stored between 1st January 1999 and 31st December 1999. We have assumed this is negligibly small.

On the basis of the estimated data in the supply column, we have total net supply of 234 mcm in 1999, or about 290 lhd using a population figure for the West Bank of 2,230,000. This average of 290 lhd is as meaningful as the average number of mothers of my grandchildren have–0.5 or one divided by two–a parallel to King Solomon's proposal. The figure of 290 combines the data for two enormously disparate abstracting groups: Palestinians and Jewish settlers.

The uses of water have next to be considered. The PWA database cross-classifies Palestinian use by source of supply (boreholes versus springs) as well as by usetype (agriculture versus other uses). Two points are worth making here. *First,* agriculture use makes up about 80 per cent of total use. *Second,* Palestinian domestic and industrial use is only 31 lhd. In an international perspective, this latter statistic is extraordinarily low. However, it should be remembered that the supply-side information does not include surface-water abstraction, the import of water from Israel, the internal reuse of wastewater and the export of water to Israel. In respect of drinking water, bottled water is probably significant in urban areas. Settler use is classified only by source. Springs predominate here because of the 100 mcm/year of brackish water it is assumed are captured in the Jordan Valley for agricultural applications.

The supply constraints faced by the West Bank population have created severe shortages in water use make it impossible to estimate accurately Palestinian domestic consumption in litres per head per day. A joint Palestinian Ministry of Agriculture, FAO and UNDP document certainly indicates allocation stress. This *Draft Strategy for Sustainable Agriculture in Palestine* refers to the water use split between agricultural and other uses, as well as the population rate of growth of nearly 4 per cent. It then suggests that non-agricultural use "is anticipated to grow exponentially.... Unless substantial additional supplies of freshwater are made available, [the West Bank and Gaza] will soon face a serious water crisis".

The point has now come when we should look at the principal characteristics of the West Bank's agriculture and, within that, the irrigated sector. It shows that rangeland (partly used for grazing) and cultivated land make up 61 per cent of the total.

However, the EIU (1999:69) reports that Israel had expropriated or closed for military reasons over 70 per cent of West Bank land by 1996, reducing the potential area available for expanding cultivation or diversifying crops. Moreover in the 6 months October 2000 to March 2001, the Israeli Army was responsible for further destruction on a wide scale. The Ministry of Agriculture classifies the West Bank into four agro-ecological regions:

- The Jordan Valley region lies 90-375 m below sea level and its climate is semi-tropical but with only 100-200 mm of rainfall per year. The soil type is predominantly *lisan marl* with soil salinization a major problem. Irrigation is essential: winter vegetables and grapes are the main crops.
- The *eastern slopes region* is a transitional zone between the Mediterranean and desert climate with annual rainfall of 150-300 mm. The main agricultural activity is livestock grazing plus small pockets of spring-irrigated cultivation. D.J. Scarpa (personal communication) notes that a significant rainfall patterns east-west. The western slopes close to the summit of the Jerusalem Hills and Hebron Mountains receive 700 mm/year. East of the summit a rain-shadow desert develops with rainfall less than 100 mm/year at the shore of the Dead Sea, a lateral distance of only 20 km.
- The *central highlands region* extends the length of the West Bank

with hills ranging from 400 to 1,020 m above sea level. Rainfall varies from 300 mm in the southern foothills to 600 mm in the north. The *terra rosa* soils are low in fertility and have a poor water-holding capacity. Agriculture is primarily rainfed, including olives, field crops, stone fruits, vegetables and forage.

- The *semi-coastal region* has an elevation of 100-300 m above sea level and rainfall of 400-700 mm/year. Its alluvial soils support rain-fed olives, stone fruits and field crops. Limited irrigation supplies are used for vegetable production.

In respect of ownership patterns, land is fragmented owing to Muslim inheritance customs. Small-scale, family-owned farms dominate agricultural production: 90 per cent of holdings are less than 5 ha in size. In the Jordan Valley region almost all major landowners are from Jericho, Nablus and Ramallah. The Wakf also owns agricultural land there.

It appears that there are no procedures in force for issuing title-deeds and hence for securing the right of land ownership or use. This stifles land's status as collateral for agricultural credit. Abundant family labour is said to account for the survival of agriculture. Nadia Farah (1997) has argued that Palestinians have historically exhibited a close affiliation with their land as part of their home life. In terms of agricultural value-added output in the West Bank and Gaza, the percentage shares in 1999 were as follows: livestock 39 per cent, vegetables 28 per cent, non-citrus fruits 11 per cent, field crops 11 per cent, olives 9 per cent, citrus fruits 5 per cent, other 7 per cent. In the West Bank alone, animal husbandry contributes 50 per cent of agricultural value-added; but olive cultivation accounts for 75 per cent of the area under trees–50 per cent of cultivated land and 25 per cent of agricultural income.

Agriculture has traditionally been a major contributor to the economy and in the 5 years 1992-96 its average annual share of overall GDP at factor cost was 27 per cent for the West Bank and Gaza; for the former, it was probably even higher. But for about the last 15 years, agriculture's share of GDP has followed a downward trend.

Crops are grown using traditional low-risk, low-input methods and productivity levels are below those of Israel but compare reasonably well with other Arab states. Specifically with respect to employment, production practices are labour intensive. Agriculture is said to have absorbed 43 per cent of the West Bank labour force in the early 1960s but this figure had dropped to about 22 per cent 30 years later. Nevertheless the *Draft Strategy* states: "Agriculture's most important economic role has been that it served as a source of livelihood for a substantial proportion of the population, as an employer of last resort, especially during occupation...."

Moreover, in considering the employment generation capacity of agriculture, we should not forget the jobs it creates indirectly in livestock and crop processing. Little rum is drunk in Palestine but, for example, the West Bank had 218 operating olive presses in 1998 employing 1,200 workers with an annual input of more than 63,000 tonnes of olives. Food processing, with metal products, textiles, clothing and leather, are the five largest components of the manufacturing sector.

Before moving off the *general* subject of agriculture, it is important to understand how difficult the sector's relations have been with Israeli institutions. The Economist Intelligence Unit writes:

> The Palestinian economy suffers from a number of structural imbalances, many of which are a direct result of Israeli occupation. Since 1967 the West Bank and Gaza Strip have been cut off from their traditional markets, with Israel the only outlet for trade. High dependence on Israel for employment and trade makes the Palestinians vulnerable to restrictions on the movement of labour and goods .... Expectations that agricultural exports would boom after Israeli markets were opened to Palestinian goods in 1994 have been dashed by the implementation of non-tariff barriers by Israel.

Prolonged Israeli closures have also raised farmers' costs and created huge surpluses, further discouraging agricultural investment. The hydrological resource is not that of a catchments–it is in contrast an area defined by political struggle, war and history. The territory includes the River Jordan as its eastern boundary, although no surface water is pumped or diverted along this stretch of the river.

The absolutely dominant sources of appropriation of water for irrigation are the wells, boreholes and springs of the mountain aquifer. Urban wastewater reuse is included, although these projects–in January 2001–had not yet come on stream. In respect of drainage, escapes do not appear; they are a surface-water phenomenon. Nor is leaching included.

It has already been shown that irrigated cultivation covers about 9,000 ha, i.e., 1.5 per cent of the West Bank's area. Further large amounts are used for agriculture on the Jewish settlements' farms.

Irrigation wells are located predominantly in the semi-coastal region from south of Qalqilya to north of Tulkarm, in the border area of the semi-coastal and central highlands regions south and north of Jenin, along the Wadi el Fara'h that bridges three regions (the central highlands, the eastern slopes and the Jordan Valley) and in the Jordan Valley east of Jericho. The principal spring discharges are between Nablus and Tubas in the central highlands and both south-west and north-west of Jericho on the Eastern Slopes.

The *Draft Strategy for Sustainable Agriculture in Palestine* provides valuable information and analysis on the West Bank's irrigation sector. The territory's climatic variability permits production of a wide range of agricultural products over the whole year. The warm winter months in the Jordan Valley are suitable for the production of vegetables and in the spring and autumn both fruit and vegetables can be grown. In the central highlands and the semi-coastal region, the moderate summer climate is also well suited to irrigated crops. The irrigated area may be very small but it contributes about one-third of agricultural output. Plastic greenhouses are also used to grow vegetables and soft fruit throughout the year. The cropping pattern for the irrigated sector in 1999, in thousands of dunums, was estimated to be:

Vegetables: 76 Field crops: 21 Citrus fruits: 16 Bananas: 6

The dunum is equal to one-tenth of a hectare. The principal vegetables grown are tomatoes, cucumbers, aubergines and squashes. The main field crops are potatoes, onions, pulses, wheat and barley. The Jordan Valley region is the chief source of vegetables (and the declining area of bananas) and the semi-coastal region is the biggest source of field crops.

The *Draft Strategy* suggests that there are about 600,000 dunums (60,000 ha) available for irrigated agriculture on the West Bank compared with the 120,000 that are presently cultivated. In part the difference is accounted for by the Israeli occupation. Before 1967, the Jordan Valley was the biggest contributor of irrigated crops. "However, after 1967, the Israeli authorities confiscated large areas next to the Jordan River for security reasons. It also destroyed about 140 irrigation pumps, which were used to pump water from the Jordan River to Palestinian farms in the Valley area."

Open-field planting is the most common method in all the agro-ecological regions for vegetables. It is highest in the Jordan Valley region because frost is less frequent there. Yields are about 0.7 tonnes/dunum for green beans to about 7 tonnes for tomatoes.

Low plastic *tunnels* cover about 20,000 dunums (2,000 ha) with average yields of about 2.8 tonnes/dunum. This is higher than open-field planting. Such tunnels are used especially in the Valley to provide protection from frost and improve the agricultural microclimate.

Plastic *houses* cover about 18,000 dunums (1,800 ha) and this total continues to grow. They allow good control for the climate and permit vegetables to be planted all year round for most areas of the West Bank. They are most common in the semi-coastal region. New vegetable varieties suitable for plastic houses have been introduced, resulting in higher yields. These may be about 9-10 tonnes/ dunum, but for crops such as cucumbers and tomatoes yields can exceed 15 tonnes/ dunum.

Irrigated fruit trees are mainly of citrus varieties in the semi-coastal region and bananas in the Jordan Valley. Bananas consume more water than citrus fruits but the 150 per cent Israeli import tariff on extra-region imports boosts bananas' market price and raises the gross margin/dunum. Particularly in the Jordan Valley, after the low rainfall of 1998-2000, the water table was dropping in 2001 and salinity increasing. Date palms are tolerant, for a temporary period, of saline content in water. But bananas require high-quality supplies. In the Valley one Palestinian farmer described how, in these conditions, he is able to continue intensive cultivation (3-5 crops per year) of banana. He captures water from the Auja spring and uses an injection well to recharge the aquifer.

Irrigation water application ranges from 400 mm in the semi-coastal region to 900 mm in the Valley and averages around 600 mm. As the reader is doubtless aware, 100 mm is equal to 100 m^3/dunum. The lower, semi-coastal application rate is attributed to the more common presence of plastic houses and the fact that water in this zone comes from reliable wells equipped with piped networks for water distribution.

The Valley's higher application rate is related to dependence on springs as the major water source. Spring discharge has high variability and large amounts of spring water are lost because of lack of storage facilities, especially in winter when use is lower. Moreover, water there is distributed by open, earth and concrete canals, with large seepage and non-beneficial evaporation losses. Extensive use of traditional irrigation methods on large areas of openfield vegetables also diminishes E_3, tonnage efficiency. The *Draft Strategy* reports an interesting development in rainwater collection for irrigation purposes:

To improve the productivity of rain-fed agriculture and create rural jobs, the PNA is supporting a land reclamation programme in selected areas by terracing the steep slopes and installing cisterns to be used in supplementary irrigation by manual water scooping. Terraces are built according to traditional techniques but supported by heavy machinery. Terraces are planted with fruit tree seedlings. Vegetables are cultivated between the trees during the first three years before water consumption of fruit trees reaches its maximum. The main fruit plantings are grapes, apples, pears, quince pomegranate, figs, peaches, almonds and plums.

To control erosion forest seedlings are planted on land adjacent to the reclaimed areas. The cost-benefit analysis of the land reclamation project shows the activity to be economically profitable with an internal rate of return of 19.8%. Building terraces and installing cisterns is a labor-intensive activity that creates employment while reclaiming the land.

Using the categorization irrigation water costs on the West Bank take two forms: own-supply costs and indirect fees. Own-supply costs are made up of pumping costs, where wells are used and the operation and maintenance of the distribution network for both wells and springs. Indirect fees are paid by farmers to well and spring share-holders for access to irrigation water usually on the basis of the length of time for which the farmer has access to a distribution channel or pipe.

Own-supply costs were as little as US$0.03/m^3 in the Valley in the year 2000. But they are said to have stood at US$0.21-0.34/m^3 in the Jenin and Tulkarem areas. So the Valley's higher application rates may also reflect the lower cost of water to its farmers. There is no implication here that the elasticity of demand for irrigation water is especially high on the West Bank. It is merely that the gradient of the demand function is sufficient to have real quantitative effect when the cost of water in one area is ten times higher than it is in another.

On the West Bank irrigation water quality is generally considered to be acceptable, with no serious indications of pollution in the deep aquifer. However, water contamination in the shallow aquifer is increasing, particularly in the more urbanized semi-coastal region. Salinization is increasing of the shallow aquifer wells in the Valley. There, farmers are resorting to mixing brackish water with fresh spring water to augment supply. Specifically with respect to the area south of Jerusalem that drains towards the Mediterranean, David Scarpa discovered, during the low rainfall event of 1998-9, that some agricultural springs exhibited faecal coliform contamination and high nitrate values. The former derives from the watering of sheep and goats, the latter from the overapplication of fertilizer. Springs for domestic use were also contaminated, such as by sulphates sourced from agriculture.

In the preceding pages we have reviewed the hydropolitical context within which any sectoral understanding of water supply and use on the West Bank must be developed. A general overview of the territory's agricultural sector has been provided as well as a review of irrigated cultivation in particular. During 1999-2000 the Food and Agriculture Organization (FAO) and the United Nations Development Programme (UNDP) carried out a joint "comparative advantage" study of the West Bank and Gaza in respect of irrigated cultivation.

Unfortunately, from the point of view of this case study, they used none of the irrigation water indicators. The efficiency indicator that is used is not defined but appears to be the ratio of crop turnover to the cost of the irrigation water used. The results for the West Bank and are easily interpreted.

Greenhouses are a winner. Fruit trees are a loser. Spring vegetables score better than those produced in the autumn. If the data are accurate, the most surprising result is that in all save the first seven ranked crops the cost of water is more than 20 per cent of total turnover. In the case of semi-coastal oranges, it is 91 per cent-an implausibly high value. In contrast, data from 1993-4 in a study prepared by the Applied Research Institute Jerusalem (1998) and entitled *Water Resources and Irrigated Agriculture in the West Bank* suggests that the cost of water as a proportion of total prime costs in the Jordan Valley averaged 11 per cent for plastic houses, 17 per cent for open-field cultivation of vegetables and 24 per cent for fruit trees. The data for northern districts of the West Bank were, respectively, 10 per cent, 19 per cent and 33 per cent.

The measure of water efficiency that is proposed as most relevant to comparison of water supplies channelled to different economic activities is sector efficiency. We do not believe that this has ever been calculated for the Palestinian economy, so it cannot be used here. As an alternative one might deploy the ratio of GDP originating in a given economic sector to the water quantities with which it is supplied. The data are clumsy but the story is clear. Irrigated agriculture in Palestine uses about 70 per cent of gross supplies of water but produces only 8 per cent of GDP. The result does not surprise but it is unmistakable.

It is certainly a feature of the sociological literature that rural life centred on a small farm, owned by those who work it, is seen as a strength of the Palestinian people and contributes substantially to their culture. In terms of employment, in the early 1990s agriculture contributed about 22 per cent of the total, compared with more than 40 per cent 30 years earlier. The *Draft Strategy* suggests" ...agriculture has served as a shock absorber for the economy in the last few years of economic recession. This also helped alleviate poverty, which would have been far worse if this avenue were not available." The unemployment rate in 1997 was about 18 per cent. But the same source suggests specific inadequacies with respect to human, social and physical capital in the irrigated sector:

In respect of capital assets, we have already looked in detail at the crux of the West Bank's problems of access to natural capital: the impossibility under the present politico-military regime of equitable use of the River Jordan and the mountain aquifer. With respect to social capital, we can draw on the fascinating study of West Bank and Gaza hydropolitics written by Julie Trottier (1999). She shows just how complex, flexible and autonomous are the irrigation water distribution networks developed at the village level by the families enjoying appropriation rights in local wells and springs.

These are based on customary law, with a basis in oral rather than written agreements. The policy implications of Trottier's grounded fieldwork, based on participant observation and participatory rural appraisal in locations such as Jericho, Artas, Falamiah, Battir and Ein Arik, contrast strongly with the low-key judgements of the *Draft Strategy*. Before leaving this brief encounter with rural livelihoods it is worth stressing that rural households derive substantial income from *urban* employment. Moreover, PASSIA (2001:265) indicates that 47,000 Palestinians with official permits and about 61,000 without, worked in Israel each day on average over the first 6 months of the year 2000. In the same year 23 per cent of the Palestinian workforce were employed in Israel. This number would have dropped sharply with the initiation of the Al Aqsa *intifada* in September 2000. The *Jerusalem Times* reported that by the end of 2000 losses due to Israel's prevention of Palestinian workers entering Israel had reached US$170 million. In 1996 remittances from Palestinians living in Israel and the Gulf were equal to US$218 million.

This can be dealt with in a few words. The West Bank and Gaza confront a situation of dire food *insecurity*. In 1999 it was possible to provide domestically 91 per cent of vegetable consumption, 90 per cent of white meat, but only 61 per cent of milk, 35 per cent of red meat and very small proportions of the basic staples wheat and rice. Agricultural imports far exceed agricultural exports and Israel is the major supplier of food and farm products. Furthermore, Palestinian merchandise trade is in chronic deficit: in 1997 imports were 5.7 times greater than exports.

This case study ends here with the strategic policy issues that face the Palestinian institutions over the next 5-10 years. It has been demonstrated that, in respect of water allocation at the regional scale, the hydrosocial baseline balance exhibits insufficient total net supply to meet the requirements of the water-using sectors. Agricultural use makes up more than 70 per cent of total use on the West Bank and so it is reasonable to explore the implications of reallocating part of that share to the domestic and industrial sectors.

In terms only of sectoral efficiency, there can be no doubt that such reallocation is justified. With respect only to rural livelihoods, irrigated cultivation plays such a vital role within the agricultural sector that reallocation cannot be justified. In respect only of food security, reallocation would make worse an already appalling situation.

So, not unexpectedly, the sectoral allocation criteria give mixed signals. Clearly the decision to reallocate or not to reallocate should be taken within the wider water resource policy framework. We suggest that strategic policy on the West Bank should be developed and implemented *simultaneously* along four tracks:

- *Water appropriation and the peace process:* Equitable access to the flows of the Jordan catchment should be agreed between the riparians. In addition, Israel and a new Palestinian state should negotiate and sign a treaty to share the flows of the mountain aquifer in a sustainable and equitable manner. It may be appropriate for the Global Water Partnership, for example, to finance an independent hydrological, engineering and social science team to facilitate these hydropolitical processes. Success on this first track would eradicate the current situation of allocation stress for the medium term.

Furthermore, in the event that the peace process between Palestine and Israel leads to the closure of the Jewish settlements on the West Bank, the PWA should take responsibility for the abandoned boreholes.

- *The West Bank's hydrosocial balance: supply:* The Palestinian institutions should invest in rainwater collection for agriculture and the urban sector. They should develop surface-water abstraction from the West Bank wadis as well as using them for aquifer recharge.

  The internal reuse of waste water by industry and households should be expanded. The key institutions should promote the external reuse of treated urban wastewater for agriculture and invest in economically viable schemes for reducing water losses in the rural and urban areas. The deterioration of existing water infrastructure such as wells, springs, ponds and canals should be reversed.
- *The West Bank's hydrosocial balance use:* Wherever it is politically feasible and where the transaction costs permit, volumetric pricing should be introduced for all the uses of water, deploying a tariff that at least covers, in the first instance, OM&M costs of supply. Such action would not be necessary for irrigated agriculture using wells, boreholes and springs, as own-supply costs already serve this purpose. Modern irrigation management practices should be stimulated. The cultivation of salt-tolerant crops should be encouraged.
- *The West Bank's water resource institutions:* The formation of water user associations should be assisted where research shows this reflects local wishes. In promoting such associations, the already existing associations autonomously developed at the village level should be recognized as archetypes. The responsibilities and powers of the Palestinian water resource institutions, including those outside the government sector, should be restructured to permit these strategic policy tracks to be effectively pursued. In the medium term the political and technical feasibility should be investigated of introducing abstraction charges to finance the water resource planning agencies.

THE RURAL AND URBAN USES OF WATER

The case study applied the method to the Palestinian West Bank. Previously we have produced such balances for the Thames catchment and for Gaza, using secondary data. To be frank, we do not see any possibility for a water resource planner to work effectively without constructing such baseline and scenario tabulations. The data for the West Bank at the present time are certainly open to question in respect of both coverage and accuracy but this first attempt to construct the balance for the West Bank can contribute to the process of improving the data sources.

It might be imagined by the unwary *must* be accurate because the data almost precisely comply with the mathematical identity of total net supply and use. The estimation error is only 6 mcm in a total net supply of 234 mcm. This conclusion would be a mistake. It is only when entirely separate calculations of the two totals are made and found to be approximately equal that that one can hope the data are good. The same was true of Thames study where, subsequent to going to press, the author discovered a gross error in the net supply estimate had been obscured precisely by the derivation of the use estimate from supply-side data.

E_{11}, sector efficiency, was recommended to be of particular value when considering water resource policy in regions where allocation stress exists. In the case of the West Bank, a primitive first estimate of the sector efficiencies of agriculture versus the urban economy gave irrigated agriculture using 70 per cent of abstractions to produce 8 per cent of GDP. But in the case of the Palestinian West Bank the gross disparity of rural and urban efficiency does not lead simply to a policy proposal for the supply switching of well and spring water to the towns. The Palestinian National Authority's most important policy drive over the next 10 years should be the augmentation of supply by securing equitable access to the Jordan River and to the mountain aquifer.

The third link between analysis and the case study that deserves comment concerns national food security. The West Bank does not enjoy food self-sufficiency nor is it capable of financing the foreign exchange required for imported food from its food exports. Neither of these facts seems to be of strategic importance. What is of real concern is that the West Bank economy is incapable of earning its food imports from manufactured exports and tourism. The heavy reliance on external assistance is unavoidable at present.

Chapter 11

Crop Evapotranspiration Under Standard Conditions

CROP EVAPOTRANSPIRATION (ET_c)

The crop coefficient approach for calculating the crop evapotranspiration under standard conditions (ET_c). The standard conditions refer to crops grown in large fields under excellent agronomic and soil water conditions. The crop evapotranspiration differs distinctly from the reference evapotranspiration (ET_o) as the ground cover, canopy properties and aerodynamic resistance of the crop are different from grass. The effects of characteristics that distinguish field crops from grass are integrated into the crop coefficient (K_c). In the crop coefficient approach, crop evapotranspiration is calculated by multiplying ET_o by K_c.

Differences in evaporation and transpiration between field crops and the reference grass surface can be integrated in a single crop coefficient (K_c) or separated into two coefficients: a basal crop (K_{cb}) and a soil evaporation coefficient (K_e), i.e., $K_c = K_{cb} + K_e$. The approach to follow should be selected as a function of the purpose of the calculation, the accuracy required and the data available.

CALCULATION PROCEDURES

Direct Calculation

The evapotranspiration rate from a cropped surface can be directly measured by the mass transfer or the energy balance method. It can also be derived from studies of the soil water balance determined from cropped fields or from lysimeters.

Crop evapotranspiration can also be derived from meteorological and crop data by means of the Penman-Monteith equation. By adjusting the albedo and the aerodynamic and canopy surface resistances to the growing characteristics of the specific crop, the evapotranspiration rate can be directly estimated. The albedo and resistances are, however, difficult to estimate accurately as they may vary continually during the growing season as climatic conditions change, as the crop develops, and with wetness of the soil surface. The canopy resistance will further be influenced by the soil water availability, and it increases strongly if the crop is subjected to water stress.

As there is still a considerable lack of consolidated information on the aerodynamic and canopy resistances for the various cropped surfaces, the FAO Penman-Monteith method is used in this handbook only for estimating ET_o, the evapotranspiration from a well-watered hypothetical grass surface having fixed crop height, albedo and surface resistance.

Crop coefficient approach [sh]

In the crop coefficient approach the crop evapotranspiration, ET_c, is calculated by multiplying the reference crop evapotranspiration, ET_o, by a crop coefficient, K_c:

$$ET_c = K_c\, ET_o$$

where ET_c crop evapotranspiration [mm d^{-1}],
K_c crop coefficient [dimensionless],
ET_o reference crop evapotranspiration [mm d^{-1}].

Most of the effects of the various weather conditions are incorporated into the ET_o estimate. Therefore, as ET_o represents an index of climatic demand, K_c varies predominately with the specific crop characteristics and only to a limited extent with climate. This enables the transfer of standard values for K_c between locations and between climates. This has been a primary reason for the global acceptance and usefulness of the crop coefficient approach and the K_c factors developed in past studies.

The reference ET_o is defined and calculated using the FAO Penman Monteith equation. The crop coefficient, K_c, is basically the ratio of the crop ET_c to the reference ET_o, and it represents an integration of the effects of four primary characteristics that distinguish the crop from reference grass. These characteristics are:

- Crop height. The crop height influences the aerodynamic resistance term, r_a, of the FAO Penman-Monteith equation and the turbulent transfer of vapour from the crop into the atmosphere. The r_a tenu appears twice in me full form of the FAO Penman-Monteith equation.
- Albedo (reflectance) of the crop-soil surface. The albedo is affected by the fraction of ground covered by vegetation and by the soil surface wetness. The albedo of the crop-soil surface influences the net radiation of the surface, R_n, which is the primary source of the energy exchange for the evaporation process.
- Canopy resistance. The resistance of me crop to vapour transfer

is affected by leaf area (number of stomata), leaf age and condition, and the degree of stomatal control. The canopy resistance influences the surface resistance, r_s.

- Evaporation from soil, especially exposed soil.

The soil surface wetness and the fraction of ground covered by vegetation influence the surface resistance, r_s. Following soil wetting, the vapour transfer rate from the soil is high, especially for crops having incomplete ground cover. The combined surface resistance of the canopy and of the soil determines the (bulk) surface resistance, r_s. The surface resistance term in the Penman-Monteith equation represents the resistance to vapour flow from within plant leaves and from beneath the soil surface.

The K_c in Equation predicts ET_c under standard conditions. This represents the upper envelope of crop evapotranspiration and represents conditions where no limitations are placed on crop growth or evapotranspiration due to water shortage, crop density, or disease, weed, insect or salinity pressures. The ET_c predicted by K_c is adjusted if necessary to non-standard conditions, $ET_{c\ adj}$, where any environmental condition or characteristic is known to have an impact on or to limit ET_c. Factors for correcting ET_c to $ET_{c\ adj}$ are described in Part C.

FACTORS DETERMINING THE CROP COEFFICIENT

The crop coefficient integrates the effect of characteristics that distinguish a typical field crop from the grass reference, which has a constant appearance and a complete ground cover. Consequently, different crops will have different K_c coefficients. The changing characteristics of the crop over the growing season also affect the K_c coefficient. Finally, as evaporation is an integrated part of crop evapotranspiration, conditions affecting soil evaporation will also have an effect on K_c.

Crop Type

Due to differences in albedo, crop height, aerodynamic properties, and leaf and stomata properties, the evapotranspiration from full grown, well-watered crops differs from ET_o.

The close spacings of plants and taller canopy height and roughness of many full grown agricultural crops cause these crops to have K_c factors that are larger than 1. The K_c factor is often 5-10% higher than the reference (where $K_c = 1.0$), and even 15-20% greater for some tall crops such as maize, sorghum or sugar cane. Typical values for the crop coefficient for full grown crops ($K_{c\ mid}$).

Crops such as pineapples, that close their stomata during the day, have very small crop coefficients. In most species, however, the stomata open as irradiance increases. In addition to the stomatal response to environment, the position and number of the stomata and the resistance of the cuticula to vapour transfer determine the water loss from the crop. Species with stomata on only the lower side of the leaf and/or large leaf resistances will have relatively smaller K_c values. This is the case for citrus and most deciduous fruit trees. Transpiration control and spacing of the trees, providing only 70% ground cover for mature trees, may cause the K_c of those trees, if cultivated without a ground cover crop, to be smaller than one.

Climate

The K_c values are typical values expected for average K_c under a standard climatic condition, which is defined as a sub-humid climate with average daytime minimum relative humidity (RH_{min}) ≈ 45% and having calm to moderate wind speeds averaging 2 m/s.

Variations in wind alter the aerodynamic resistance of the crops and hence their crop coefficients, especially for those crops that are substantially taller than the hypothetical grass reference. The effect of the difference in aerodynamic properties between the grass reference surface and agricultural crops is not only crop specific.

It also varies with the climatic conditions and crop height. Because aerodynamic properties are greater for many agricultural crops as compared to the grass reference, the ratio of ET_c to ET_o (i.e., K_c) for many crops increases as wind speed increases and as relative humidity decreases. More arid climates and conditions of greater wind speed will have higher values for K_c. More humid climates and conditions of lower wind speed will have lower values for K_c.

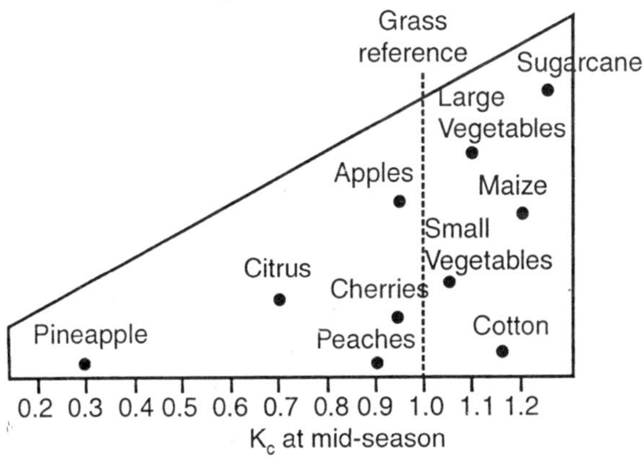

Fig. Typical K_c for Different Types of Full Grown Crops

The relative impact of the climate on K_c for full grown crops. The upper bounds represent extremely arid and windy conditions, while the lower bounds are valid under very humid and calm weather conditions.

The ranges expected in K_c as climate and weather conditions change are quite small for short crops but are large for tall crops. Guidelines for the adjustment of K_c to the climatic conditions as a function of crop height.

Under humid and calm wind conditions, K_c becomes less dependent on the differences between the aerodynamic components of ET_c and ET_o and the K_c values for 'full-cover' agricultural crops do not exceed 1.0 by more than about 0.05. This is because full-cover agricultural crops and the reference crop of clipped grass both provide for nearly maximum absorption of shortwave radiation, which is the primary energy source for evaporation under humid and calm conditions.

Generally, the albedos, α, are similar over a wide range of full-cover agricultural crops, including the reference crop. Because the vapour pressure deficit ($e_s - e_a$) is small under humid conditions, differences in ET caused by differences in aerodynamic resistance, r_a, between the agricultural crop and the reference crop are also small, especially with low to moderate wind speed.

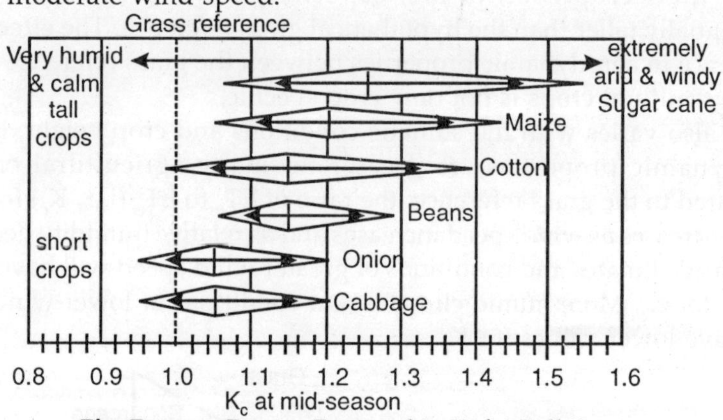

Fig. Extreme Ranges Expected in K_c for Full Grown Crops as Climate and Weather Change

Under arid conditions, the effect of differences in r_a between the agricultural crop and the grass reference crop on ET_c become more pronounced because the ($e_s - e_a$) term may be relatively large. The larger magnitudes of ($e_s - e_a$) amplify differences in the aerodynamic term in the numerator of the Penman-Monteith equation for both the crop and the reference crop. Hence, K_c will be larger under arid conditions when the agricultural crop has a leaf area and roughness height that are greater than that of the grass reference.

Because the $1/r_a$ term in the numerator of the Penman-Monteith equation is multiplied by the vapour pressure deficit ($e_s - e_a$), the ET from tall crops increases proportionately more relative to ET_o than does ET from short crops when relative humidity is low. The K_c for tall crops, such as those 2-3 m in height, can be as much as 30% higher in a windy, arid climate as compared with a calm, humid climate. The increase in K_c is due to the influence of the larger aerodynamic roughness of the tall crop relative to grass on the transport of water vapour from the surface.

SOIL EVAPORATION

Differences in soil evaporation and crop transpiration between field crops and the reference surface are integrated within the crop coefficient. The K_c coefficient for full-cover crops primarily reflects differences in transpiration as the contribution of soil evaporation is relatively small. After rainfall or irrigation, the effect of evaporation is predominant when the crop is small and scarcely shades the ground.

For such low-cover conditions, the K_c coefficient is determined largely by the frequency with which the soil surface is wetted. Where the soil is wet for most of the time from irrigation or rain, the evaporation from the soil surface will be considerable and K_c may exceed 1. On the other hand, where the soil surface is dry, evaporation is restricted and K_c will be small and might even drop to as low as 0.1. Differences in soil evaporation between the field crop and the reference surface can be forecast more precisely by using a dual crop coefficient.

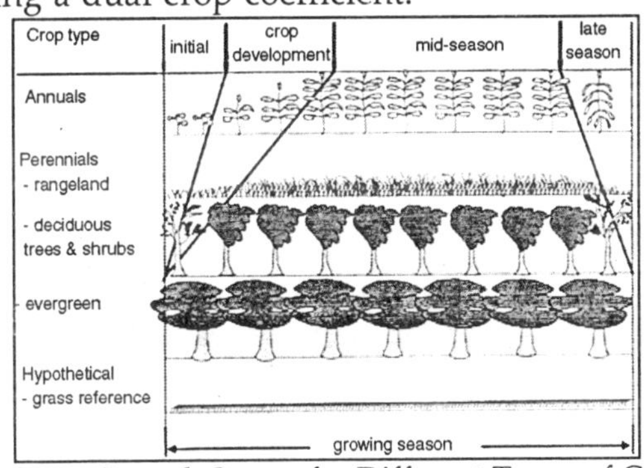

Fig. Crop Growth Stages for Different Types of Crops

CROP GROWTH STAGES

As the crop develops, the ground cover, crop height and the leaf area change. Due to differences in evapotranspiration during the various growth stages, the K_c for a given crop will vary over the growing period. The growing period can be divided into four distinct growth stages: initial, crop development, mid-season and late season.

Initial Stage

The initial stage runs from planting date to approximately 10% ground cover. The length of the initial period is highly dependent on the crop, the crop variety, the planting date and the climate. The end of the initial period is determined as the time when approximately 10% of the ground surface is covered by green vegetation. For perennial crops, the planting date is replaced by the 'greenup' date, i.e., the time when the initiation of new leaves occurs.

During the initial period, the leaf area is small, and evapotranspiration is predominately in the form of soil evaporation. Therefore, the K_c during the initial period ($K_{c\ ini}$) is large when the soil is wet from irrigation and rainfall and is low when the soil surface is dry. The time for the soil surface to dry is determined by the time interval between wetting events, the evaporation power of the atmosphere (ET_0) and the importance of the wetting event. General estimates for $K_{c\ ini}$ as a function of the frequency of wetting and ET_0 are given in Table. The data assume a medium textured soil. The procedure for estimating $K_{c\ ini}$ is presented.

Table. Approximate Values for $K_{c\ ini}$ for Medium Wetting Events (10-40 mm) and a Medium Textured Soil

Wetting Interval	*Evaporating Power of the Atmosphere (ETo)*			
	Low 1-3 mm/day	Moderate 3-5 mm/day	High 5-7 mm/day	Very High > 7 mm/day
Less than weekly	1.2-0.8	1.1-0.6	1.0-0.4	0.9-0.3
Weekly	0.8	0.6	0.4	0.3
Longer than once Per week	0.7 - 0.4	0.4 - 0.2*	0.3 - 0.2*	0.2*- 0.1*

Crop Development Stage

The crop development stage runs from 10% ground cover to effective full cover. Effective full cover for many crops occurs at the initiation of flowering. For row crops where rows commonly interlock leaves such as beans, sugar beets, potatoes and corn, effective cover can be defined as the time when some leaves of plants in adjacent rows begin to intermingle so that soil shading becomes nearly complete, or when plants reach nearly full size if no intermingling occurs.

For some crops, especially those taller than 0.5 m, the average fraction of the ground surface covered by vegetation (f_c) at the start of effective full cover is about 0.7-0.8. Fractions of sunlit and shaded soil and leaves do not change significantly with further growth of the crop beyond $f_c \approx$ 0.7 to 0.8. It is understood that the crop or plant can continue to grow in both height and leaf area after the time of effective full cover. Because it is difficult to visually determine when densely sown vegetation such as winter and spring cereals and some grasses reach effective full cover, the more easily detectable stage of heading (flowering) is generally used for these types of crops. For dense grasses, effective full cover may occur at about 0.10-0.15 m height. For thin stands of grass (dry rangeland), grass height may approach 0.3-0.5 m before effective full cover is reached. Densely planted forages such as alfalfa and clover reach effective full cover at about 0.3-0.4 m.

Another way to estimate the occurrence of effective full cover is when the leaf area index (LAI) reaches three. LAI is defined as the average total area of leaves (one side) per unit area of ground surface.

As the crop develops and shades more and more of the ground, evaporation becomes more restricted and transpiration gradually becomes the major process. During the crop development stage, the K_c value corresponds to amounts of ground cover and plant development. Typically, if the soil surface is dry, $K_c = 0.5$ corresponds to about 25-40% of the ground surface covered by vegetation due to the effects of shading and due to microscale transport of sensible heat from the soil into the vegetation.

A $K_c = 0.7$ often corresponds to about 40-60% ground cover. These values will vary, depending on the crop, frequency of wetting and whether the crop uses more water than the reference crop at full ground cover (e.g., depending on its canopy architecture and crop height relative to clipped grass).

Mid-season Stage

The mid-season stage runs from effective full cover to the start of maturity. The start of maturity is often indicated by the beginning of the ageing, yellowing or senescence of leaves, leaf drop, or the browning of fruit to the degree that the crop evapotranspiration is reduced relative to the reference ET_o. The mid-season stage is the longest stage for perennials and for many annuals, but it may be relatively short for vegetable crops that are harvested fresh for their green vegetation.

At the mid-season stage the K_c reaches its maximum value. The value for K_c ($K_{c\ mid}$) is relatively constant for most growing and cultural conditions. Deviation of the $K_{c\ mid}$ from the reference value '1' is primarily due to differences in crop height and resistance between the grass reference surface and the agricultural crop and weather conditions.

Late Season Stage

The late season stage runs from the start of maturity to harvest or full senescence. The calculation for K_c and ET_c is presumed to end when the crop is harvested, dries out naturally, reaches full senescence, or experiences leaf drop.

For some perennial vegetation in frost free climates, crops may grow year round so that the date of termination may be taken as the same as the date of 'planting'.

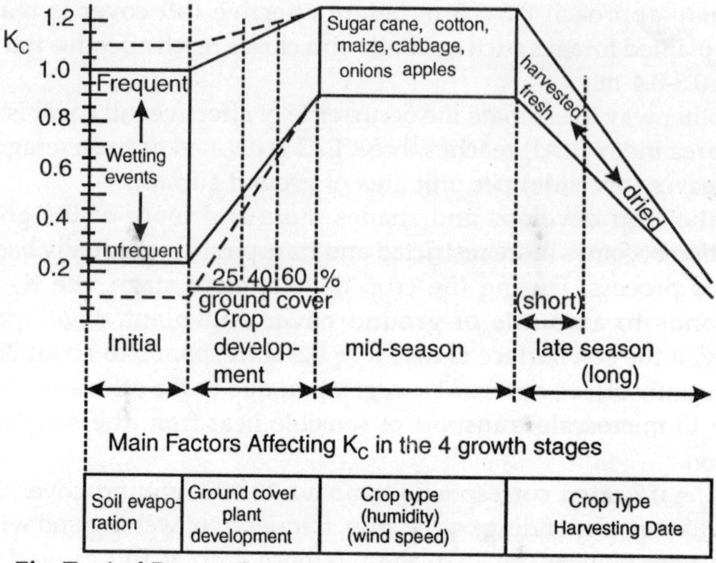

Fig. Typical Ranges Expected in K_c for the Four Growth Stages

The K_c value at the end of the late season stage ($K_{c\ end}$) reflects crop and water management practices. The $K_{c\ end}$ value is high if the crop is frequently irrigated until harvested fresh. If the crop is allowed to senesce and to dry out in the field before harvest, the $K_{c\ end}$ value will be small. Senescence is usually associated with less efficient stomatal conductance of leaf surfaces due to the effects of ageing, thereby causing a reduction in K_c.

SINGLE AND DUAL CROP COEFFICIENT APPROACHES

Single Crop Coefficient Approach (K_c)

In the single crop coefficient approach, the effect of crop transpiration and soil evaporation are combined into a single K_c coefficient. The coefficient integrates differences in the soil evaporation and crop transpiration rate between the crop and the grass reference surface. As soil evaporation may fluctuate daily as a result of rainfall or irrigation, the single crop coefficient expresses only the time-averaged (multi-day) effects of crop evapotranspiration.

As the single K_c coefficient averages soil evaporation and transpiration, the approach is used to compute ET_c for weekly or longer time periods, although calculations may proceed on a daily time step. The time-averaged single K_c is used for planning studies and irrigation system design where the averaged effects of soil wetting are acceptable and relevant. This is the case for surface irrigation and set sprinkler systems where the time interval between successive irrigation is of several days, often ten days or more. For typical irrigation management, the time-averaged single K_c is valid.

Dual Crop Coefficient Approach *($K_{cb} + K_e$)*

In the dual crop coefficient approach, the effects of crop transpiration and soil evaporation are determined separately. Two coefficients are used: the basal crop coefficient (K_{cb}) to describe plant transpiration, and the soil water evaporation coefficient (K_e) to describe evaporation from the soil surface. The single K_c coefficient is replaced by:

$$K_c = K_{cb} + K_e$$

where K_{cb} basal crop coefficient,
K_e soil water evaporation coefficient.

The basal crop coefficient, K_{cb}, is defined as the ratio of ET_c to ET_o when the soil surface layer is dry but where the average soil water content of the root zone is adequate to sustain full plant transpiration. The K_{cb} represents the baseline potential K_c in the absence of the additional effects of soil wetting by irrigation or precipitation. The soil evaporation coefficient, K_e, describes the evaporation component from the soil surface. If the soil is wet following rain or irrigation, K_e may be large. However, the sum of K_{cb} and K_e can never exceed a maximum value, $K_{c\ max}$, determined by the energy available for evapotranspiration at the soil surface.

As the soil surface becomes drier, K_e becomes smaller and falls to zero when no water is left for evaporation. The estimation of K_e requires a daily water balance computation for the calculation of the soil water content remaining in the upper topsoil.

The dual coefficient approach requires more numerical calculations than the procedure using the single time-averaged K_c coefficient. The dual procedure is best for real time irrigation scheduling, for soil water balance computations, and for research studies where effects of day-to-day variations in soil surface wetness and the resulting impacts on daily ET_c, the soil water profile, and deep percolation fluxes are important. This is the case for high frequency irrigation with microirrigation systems or lateral move systems such as centre pivots and linear move systems.

CROP COEFFICIENT CURVE

After the selection of the calculation approach, the determination of the lengths for the crop growth stages and the corresponding crop coefficients, a crop coefficient curve can be constructed. The curve represents the changes in the crop coefficient over the length of the growing season. The shape of the curve represents the changes in the vegetation and ground cover during plant development and maturation that affect the ratio of ET_c to ET_o. From the curve, the K_c factor and hence ET_c can be derived for any period within the growing season.

Single Crop Coefficient

Shortly after the planting of annuals or shortly after the initiation of new leaves for perennials, the value for K_c is small, often less than 0.4. The K_c begins to increase from the initial K_c value, $K_{c\ ini}$, at the beginning of rapid plant development and reaches a maximum value, $K_{c\ mid}$, at the time of maximum or near maximum plant development. During the late season period, as leaves begin to age and senesce due to natural or cultural practices, the K_c begins to decrease until it reaches a lower value at the end of the growing period equal to $K_{c\ end}$.

Dual Crop Coefficient

The single 'time-averaged' K_c curve incorporates averaged wetting effects into the K_c factor. The value for $K_{c\ mid}$ is relatively constant for most growing and cultural conditions. However, the values for $K_{c\ ini}$ and $K_{c\ end}$ can vary considerably on a daily basis, depending on the frequency of wetting by irrigation and rainfall. The dual crop coefficient approach calculates the actual increases in K_c for each day as a function of plant development and the wetness of the soil surface.

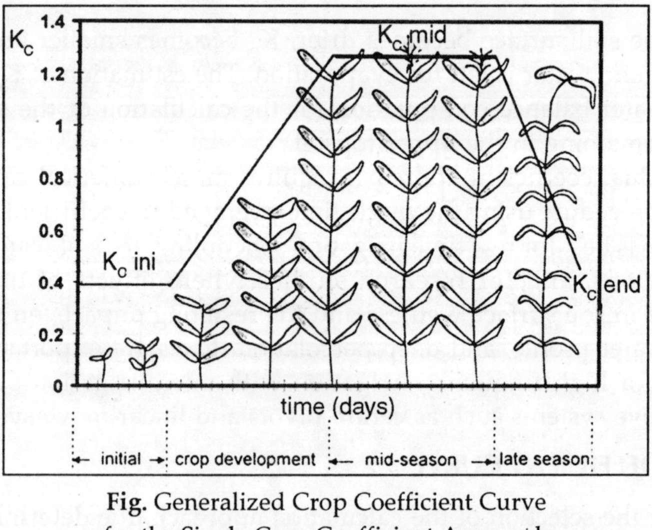

Fig. Generalized Crop Coefficient Curve for the Single Crop Coefficient Approach

As the single K_c coefficient includes averaged effects of evaporation from the soil, the basal crop coefficient, K_{cb} describing only plant transpiration, lies below the K_c value. The largest difference between K_c and K_{cb} is found in the initial growth stage where evapotranspiration is predominantly in the form of soil evaporation and crop transpiration is still small. Because crop canopies are near or at full ground cover during the mid-season stage, soil evaporation beneath the canopy has less effect on crop evapotranspiration and the value for K_{cb} in the mid-season stage will be nearly the same as K_c.

Depending on the ground cover, the basal crop coefficient during the mid-season may be only 0.05-0.10 lower than the K_c value. Depending on the frequency with which the crop is irrigated during the late season stage, K_{cb} will be similar to (if infrequently irrigated) or less than the K_c value.

Typical shapes for the K_{cb}, K_e and single K_c curves. The K_{cb} curve in the figure represents the minimum K_c for conditions of adequate soil water and dry soil surface. The K_e 'spikes' in the figure represent increased evaporation when precipitation or irrigation has wetted the soil surface and has temporarily increased total ET_c. These wet soil evaporation spikes decrease as the soil surface layer dries. The spikes generally reach a maximum value of 1.0-1.2, depending on the climate, the magnitude of the wetting event and the portion of soil surface wetted.

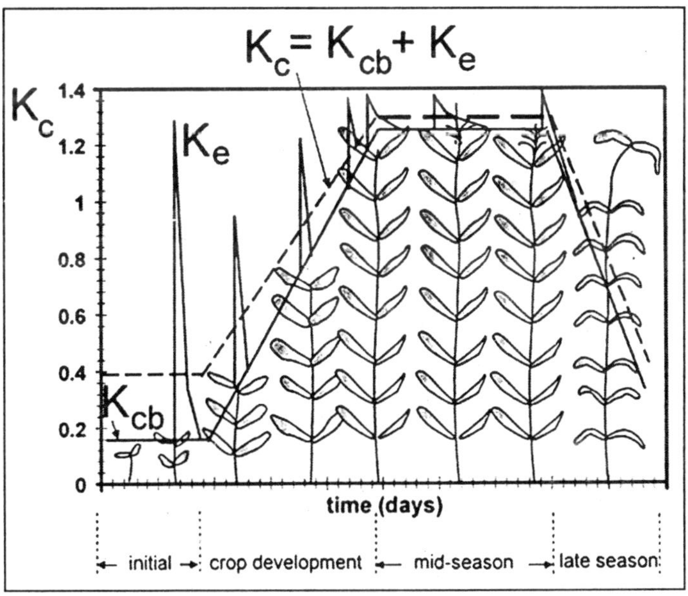

Fig. Crop Coefficient Curves Showing the Basal K_{cb} (Thick Line), Soil Evaporation K_e (Thin Line) and the Corresponding Single $K_c = K_{cb} + K_e$ Curve (Dashed Line)

Summed together, the values for K_{cb} and for K_e represent the single crop coefficient, K_c. The total K_c curve the effect of averaging $K_{cb} + K_e$ over time and is displayed as a 'smoothed' curve. It is this smoothed curve that is represented by the single K_c calculation procedure. The K_c curve lies above the K_{cb} curve, with potentially large differences during the initial and development stages, depending on the frequency of soil wetting.

Flow Chart of the Calculations

A single time-averaged crop coefficient is used to calculate ET_c. The approach using two coefficients to describe the effects of crop and soil.

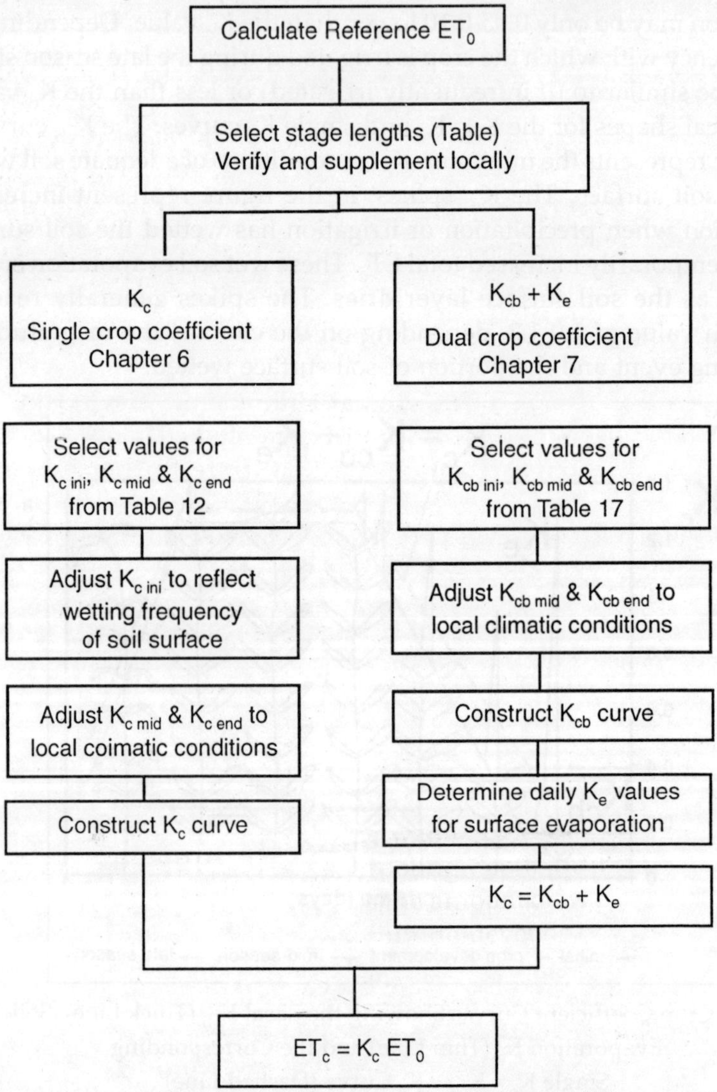

Fig. General Procedure for Calculating ET_c

SOIL WATER

At the land surface incoming precipitation, throughfall, stemflow, leafdrip, crowndrip and runoff can either infiltrate into the soil matrix or remain at the surface, collecting as detention storage and then running off through a drainage network.

The fundamental control on this 'hydrological crossroads' is the relationship between the rate at which water is received at the surface and the rate at which water is able to infiltrate the soil pore matrix. The rate of water receipt at a point depends upon the intensity of precipitation receipt and the magnitude of any precipitation convergence through canopy interception or surface runoff processes along with any lateral flows from upslope/upstream. The rate of infiltration is determined by: the available water, the characteristics of the soil surface and the exact nature of the soil's network of pores.

This hydrological crossroads is critical since it determines whether incoming water follows a rapid land surface route through the land-based components of the hydrological cycle or a much slower subsurface route, watering vegetation and recharging aquifers en route. The route has implications for the type of water resource available, the maintenance of 'ecological flows' in rivers and the potential for hydrological hazards (floods and the like).

Even where water enters the soil as infiltration, it may return to the surface further downslope – but it will certainly be changed in timing and properties. Similarly, runoff may infiltrate further downslope (therefore the hydrological crossroads that occurs where rainfall/ throughfall hits the soil surface is not the only opportunity for water exchange across the air–soil boundary).

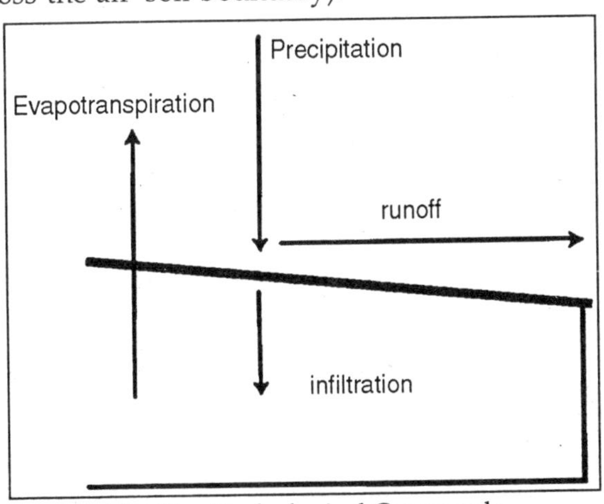

Fig. The Hydrological Crossroads

PROPERTIES OF THE SOIL

Soil is a three-phase medium comprising mineral and organic solids as well as voids filled with air and/or water. One may differentiate the mineral solids according to their size into clays (< 0.002 mm), silts (0.002–0.063 mm), sands (0.063–2 mm) and rock fragments (> 2 mm). The size distribution of solids is important since it determines the size distribution of pores in the soil, which determines the hydraulic properties of the soil. The mineral components of a soil are derived from the parent material, so their properties (density, particle size, hardness, chemical composition) are a function of:

- The nature of the parent material.
- The period and intensity of weathering.

Soil particles tend to have a density of between 2.25 g/cm^3 (limestone) and 2.6 g/cm^3 (quartz), reflecting the range of densities of parent materials, and thus they have very little or no internal porespace. Since these particles are of random size and shape they do not tessellate (fit together) well, so that soils are a loosely held porous matrix with a bulk density closer to 1.0 g/cm^3. The porosity of a soil can be calculated from knowledge of its particle and bulk densities as:

$$P = 1 - BD/PD$$

where P = soil porosity
BD = soil bulk density
PD = soil particle density.

For the examples outlined above, this equation produces porosities of 0.62 (quartz) and 0.55 (limestone) with a bulk density of 1.0. Soils are commonly 60 per cent porespace, and it is into this porespace that surface water infiltrates and soil moisture accumulates. The soil water balance can be described using the following equation:

$$\Delta SMC = f + c - d - ET + TFLO_{in} - TFLO_{out}$$

where

ΔSMC = change in soil moisture

f = infiltration rate

c = rate of addition of water added by capillary action from the water table

d = rate of drainage (recharge) of water to the water table

ET = evapotranspiration losses

$TFLO$ = lateral subsurface seepage (throughflow).

Where inputs are greater than losses, soil moisture increases, and where losses are greater than inputs, soil moisture decreases. Soil moisture is usually measured as a volumetric fraction (m^3 water/m3soil) but on occasion the pore fraction (m^{3}water/m^3 porespace) is used. Since many of the fluxes in and out of a soil profile are driven by soil moisture, there is a dynamic feedback between the rates of infiltration, evapotranspiration, throughflow and recharge and the moisture content of the soil.

INFILTRATION

Infiltration is defined as the transfer of water across the air–soil interface or into the soil porespace. The original work on the theory of infiltration was carried out by R.E. Horton at the beginning of the twentieth century. Infiltration models are still being refined, in particular to account for spatial variability, but also to account for infiltration into very large (macro) pores. Infiltration rates represent the volumetric rate of seepage into the upper surface of a soil expressed in mm3/mm2 of soil surface (in other words, in mm depth). Rates are both spatially and temporally variable and tend to decline throughout a storm as the easily accessible porespace is filled. Infiltration capacity is a much more stable measure and may be treated as a constant for a particular soil type. Infiltration capacity is the maximum rate at which a soil can absorb water under a particular condition. The most commonly used condition is soil saturation, which yields the saturated infiltration capacity.

Controls on Infiltration

Saturated infiltration capacity varies with soil properties, vegetation, soil faunal activities, land management, slope, geomorphic location and even temperature. The most important soil properties are bulk density (porosity), texture and crumb structure. Vegetation can assist infiltration through:

- The addition of organic matter, which encourages good soil structure and low bulk density.
- Attracting soil fauna which serve to break up and aerate dense soil clods.
- Providing channels for rapid infiltration along roots.
- Protecting the soil surface from sealing through raindrop impact.

Soil faunal activities both break up dense soil clods and aerate them, but also encourage better soil structure by binding particulates together with mucus deposits. The sheer magnitude of soil faunal populations can make this effect a significant one. Land management practices affect soil properties in a number of ways, either compacting soils (e.g. grazing) or dispersing them (e.g. ploughing, tillage). Given that these practices can be carried out mechanically over very large areas, significant (catchment scale) impacts on runoff ratios (runoff per unit rainfall) can result. Slope gradient determines the opportunity time for infiltration. Where slopes are steep, runoff is generated rapidly and there is little time for seepage into the surface. For this reason, where water is limited, slopes are terraced to increase the opportunity time for infiltration. The position of a surface within the drainage network (its geomorphic location) determines the amounts of surface and subsurface water available and thus the infiltration capacity.

MEASURING INFILTRATION

Infiltration Rings

Infiltration is measured using a range of different techniques. The simplest device is the infiltration ring or infiltrometer, which is either a single or dual cylinder inserted into the soil and filled with water. Infiltration is measured as the rate of decline in the head of water. Dual ring devices avoid the problem of overestimation because of lateral dispersion of infiltration water, since the outer ring is filled first to create a circular shaft of infiltrated water and measurements are taken on the falling head of the inner ring only. The main problems with infiltration rings are that:

- Insertion leads to damage of the soil surface, especially where a crumb is present.
- The infiltration rates are for ponded infiltration only and infiltration occurs under a varying weight (or head) of water which can lead to overestimation, though constant head infiltrometers rectify this problem to some extent.
- Measurements are taken over very small areas which means that many replicates are required in a spatially variable environment.

Rainfall Simulation

An alternative to the use of infiltration rings is to set up a rainfall simulator or set of nozzles simulating rainfall over a specific field area normally between 0.5 and 20 m2. Simulation experiments are carried out over bounded plots, and infiltration is calculated as the input rainfall minus the measured overland flow. This provides a more spatially representative measure, which is less prone to experimental error but is more costly and challenging to set up and less portable.

Plot and Catchment

Since infiltration characteristics vary with characteristics of the rainfall as well as seasonal changes to the vegetation, soil and surface, infiltration rings and rainfall simulation can only really be used to calculate standard infiltration properties such as saturated infiltration capacity and infiltration rate at a specific soil moisture content. Monitored overland flow plots, in which rainfall and overland flow are continuously monitored on a bounded plot, can be used to better understand the response of infiltration over a particular area to changing (and realistic) rainfall characteristics and to seasonal variability in soil and surface properties.

Since real rainfall is used rather than simulated rainfall, these plots can be much larger and can either be bounded or make use of natural topographic boundaries to cover larger areas. In some cases, runoff measured from whole catchments in streams and rivers has been used to understand catchment scale infiltration properties, but in these cases the relationship is complicated by interception losses and evaporation.

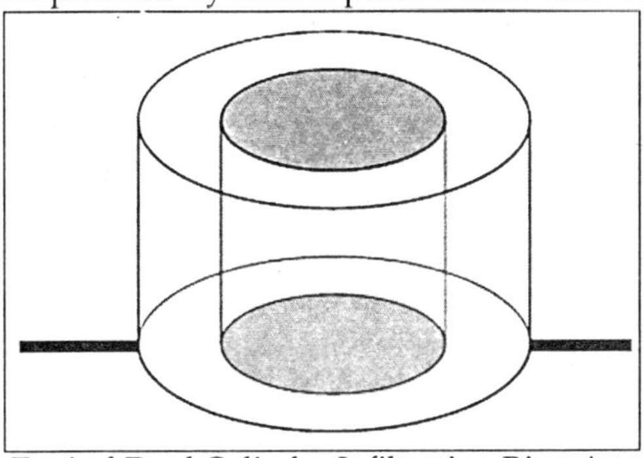

Fig. A Typical Dual Cylinder Infiltration Ring Apparatus

STORAGE OF SOIL WATER

Soil water is considered to exist in a zone of aeration in which pores are predominantly filled with air and obtain water through infiltration from the surface while losing it by drainage to the saturated zone below. The upper surface of the saturated zone is known as the water table. The saturated zone loses water to groundwater and to stream banks, a process known as effluent seepage.

This supplies baseflow to rivers and streams. In reality, soils can contain a number of perched water tables above areas or layers of low conductivity or permeability. Soil moisture refers to water within the zone of aeration; if this water fluxes downwards it becomes drainage or recharge and if it fluxes laterally it becomes throughflow. The maximum amount of water that a soil can hold in a free-draining situation is known as the field capacity, and the soil moisture deficit (smd) is the amount of water needed to raise the soil moisture content to field capacity. At the other end of the water content spectrum, wilting point is the moisture content at which plants exhibit permanent wilting (i.e. the suction that is required to release the remaining soil water from the particles is too great for the plants to exert).

Moisture is held within soils as a result of absorptive, adsorptive, osmotic and capillary forces. Absorption involves the uptake of moisture into a body. Clay minerals absorb water into their crystalline structure and swell; this water is not normally available for uptake by plants and is lost only to direct evaporation. Adsorption involves a thin film of water molecules on the surface of particles held either by weak electrical charges or by the sharing of electrons. Osmotic forces occur where water is drawn into the soil aggregates by a saline solution. Capillary forces dominate and are due to surface tension in fine pores. Their strength increases as the pore diameter falls. The total force required to overcome all of these forces and remove water from soil is known as the matric potential (a suction that must be exerted to remove water from the soil surface). Matric potential increases with decreasing soil moisture content and rises very rapidly at low water content. A saturated soil matrix with only the macropores drained would have a matric potential of zero.

Drier soils have a matric potential of less than zero and are said to be under tension. In a fully saturated soil with all pores saturated, there will be a positive pore water pressure (a matric pressure). The relationship between soil moisture and matric potential is primarily controlled by the size distribution of pores and the size distribution of particles (through capillary forces). This relationship is known as the soil moisture characteristic.

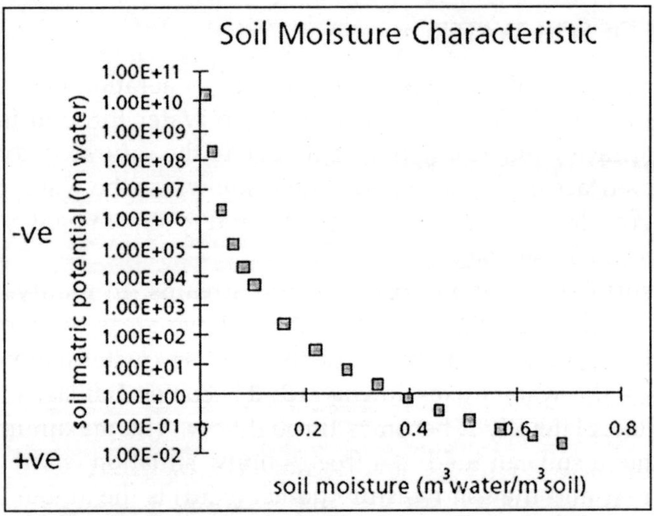

Fig. An Example of a Soil Moisture Characteristic (Soil Moisture vs. Matric Potential) for a Clay Soil.

MOVEMENT OF SOIL WATER

Throughflow is the dominant lateral flow of water in soils and can be divided into slow diffuse flow and rapid macropore flow. Diffuse flow occurs in response to a gradient in potential known as a head. The total soil water potential is the sum of the matric potential and the gravitational potential (the elevation of the point above some reference datum). Soil water migrates from areas of high potential to areas of lower potential. On sloping land, the gravitational potential dominates, so water flows downslope. As a result, soil water is distributed according to topography. Locally, matric potential gradients may dominate and water may, for example, move upwards through a profile if the surface is much drier than the subsurface. The movement of water through a porous matrix is described by Darcy's Law, which states that the discharge of water through a cross-section of saturated soil with area A is given as:

$$Q = KIA$$

where Q = discharge

K = the saturated hydraulic conductivity (rate of transmission of water) of the soil

I = the gradient in total soil water potential in the direction of flow (the hydraulic gradient).

The Richards equation expands Darcy's Law to apply to non-saturated soil conditions by allowing the hydraulic conductivity to vary with water content, through a set of complex, non-linear partial differentials. Water flows most readily through existing films of water so the hydraulic conductivity, partly a function of the distribution of pore sizes and porespace tortuosity, increases as soil moisture increases, reaching a maximum value – the saturated hydraulic conductivity or K_{sat} at pore saturation.

SPATIAL PATTERNS OF SOIL MOISTURE

At the broadest scale, soil moisture patterns are dominated by the balance between precipitation and evapotranspiration – in other words, by largely climatic variables. At the landscape scale, the dominant control becomes topographical position and its control on gravitational potential. One can calculate gravitational potential from change in height and draw flow directions, assuming that water travels from its original point to the neighbouring point with the steepest drop in altitude. Realised within a geographical information system (GIS). This shows a local drainage direction map, calculated for the digital elevation model (DEM) for a tropical catchment in Colombia which has been digitised into 25 m resolution grid cells. It shows three types of drainage area:

(a) A laminar or planar slope, in which the area contributing flow increases proportionally with distance downslope.

(b) A convergence zone (hollow), in which the area contributing increases more than proportionally with distance downslope.

(c) An area of divergence (spur), where the contributing area increases less than proportionally with distance downslope.

Relative soil wetness at a point in the landscape can be estimated according to the balance between upslope area contributing surface and subsurface flow to that point – the upslope area (A) and the potential of that point to shed water proportional to its local slope (β). Where contributing areas are high and local slopes low, soils will be relatively wet and conversely, where contributing areas are low and slopes high, soils will be relatively dry. This notion is formalised in the TOPMODEL equation:

$$WI = ln(A/\tan\beta)$$

where WI = top model wetness index (a relative measure of local wetness)

Ln = the natural logarithm.

Fig. The Local Drainage Network for a Colombian Catchment with (a) Planar, (b) Divergent and (c) Convergent Flow Areas Indicated

ET_c UNDER SOIL WATER STRESS CONDITIONS

Forces acting on the soil water decrease its potential energy and make it less available for plant root extraction. When the soil is wet, the water has a high potential energy, is relatively free to move and is easily taken up by the plant roots. In dry soils, the water has a low potential energy and is strongly bound by capillary and absorptive forces to the soil matrix, and is less easily extracted by the crop.

When the potential energy of the soil water drops below a threshold value, the crop is said to be water stressed. The effects of soil water stress are described by multiplying the basal crop coefficient by the water stress coefficient, K_s:

$$ET_{c\ adj} = (K_s K_{cb} + K_e) ET_o$$

For soil water limiting conditions, $K_s < 1$. Where there is no soil water stress, $K_s = 1$.

K_s describes the effect of water stress on crop transpiration. Where the single crop coefficient is used, the effect of water stress is incorporated into K_c as:

$$ET_{c\ adj} = K_s\ K_c\ ET_o$$

Because the water stress coefficient impacts only crop transpiration, rather than evaporation from soil, the application of K_s using Equation is generally more valid than is application using Equation. However, in situations where evaporation from soil is not a large component of ET_c, use of Equation will provide reasonable results.

SOIL WATER AVAILABILITY

Total Available Water (TAW)

Soil water availability refers to the capacity of a soil to retain water available to plants. After heavy rainfall or irrigation, the soil will drain until field capacity is reached. Field capacity is the amount of water that a well-drained soil should hold against gravitational forces, or the amount of water remaining when downward drainage has markedly decreased. In the absence of water supply, the water content in the root zone decreases as a result of water uptake by the crop. As water uptake progresses, the remaining water is held to the soil particles with greater force, lowering its potential energy and making it more difficult for the plant to extract it. Eventually, a point is reached where the crop can no longer extract the remaining water. The water uptake becomes zero when wilting point is reached. Wilting point is the water content at which plants will permanently wilt.

As the water content above field capacity cannot be held against the forces of gravity and will drain and as the water content below wilting point cannot be extracted by plant roots, the total available water in the root zone is the difference between the water content at field capacity and wilting point:

$$TAW = 1000(\theta_{FC} - \theta_{WP})Z_r$$

where TAW the total available soil water in the root zone [mm],

θ_{FC} the water content at field capacity [$m^3\ m^{-3}$],

θ_{WP} the water content at wilting point [$m^3\ m^{-3}$],

Z_r the rooting depth [m].

TAW is the amount of water that a crop can extract from its root zone, and its magnitude depends on the type of soil and the rooting depth.

Readily Available Water (RAW)

Although water is theoretically available until wilting point, crop water uptake is reduced well before wilting point is reached. Where the soil is sufficiently wet, the soil supplies water fast enough to meet the atmospheric demand of the crop, and water uptake equals ET_c. As the soil water content decreases, water becomes more strongly bound to the soil matrix and is more difficult to extract. When the soil water content drops below a threshold value, soil water can no longer be transported quickly enough towards the roots to respond to the transpiration demand and the crop begins to experience stress. The fraction of TAW that a crop can extract from the root zone without suffering water stress is the readily available soil water:

$$RAW = p\ TAW$$

where

RAW the readily available soil water in the root zone [mm],

p average fraction of Total Available Soil Water (TAW) that can be depleted from the root zone before moisture stress (reduction in ET) occurs [0-1].

The factor p differs from one crop to another. The factor p normally varies from 0.30 for shallow rooted plants at high rates of ET_c (> 8 mm d^{-1}) to 0.70 for deep rooted plants at low rates of ET_c (< 3 mm d^{-1}). A value of 0.50 for p is commonly used for many crops.

The fraction p is a function of the evaporation power of the atmosphere. At low rates of ET_c, the p values are higher than at high rates of ET_c. For hot dry weather conditions, where ET_c is high, p is 10-25% less than the values presented, and the soil is still relatively wet when the stress starts to occur. When the crop evapotranspiration is low, p will be up to 20% more than the listed values. Often, a constant value is used for p for a specific growing period, rather than varying the value each day. A numerical approximation for adjusting p for ET_c rate is $p = p_{Table\ 22} + 0.04\ (5 - ET_c)$ where the adjusted p is limited to $0.1 \leq p \leq 0.8$ and ET_c is in mm/day.

Water Stress Coefficient (K_s)

The effects of soil water stress on crop ET are described by reducing the value for the crop coefficient. This is accomplished by multiplying the crop coefficient by the water stress coefficient. K_s.

Water content in me root zone can also be expressed by root zone depletion, D_r, i.e., water shortage relative to field capacity. At field capacity, the root zone depletion is zero ($D_r = 0$). When soil water is extracted by evapotranspiration, the depletion increases and stress will be induced when D_r becomes equal to RAW. After the root zone depletion exceeds RAW (the water content drops below the threshold θ_t), the root zone depletion is high enough to limit evapotranspiration to less than potential values and the crop evapotranspiration begins to decrease in proportion to the amount of water remaining in the root zone.

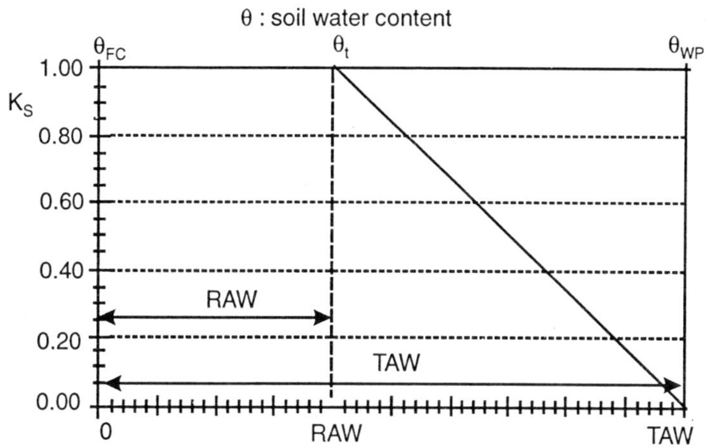

Fig. Water Stress Coefficient, K_s

SOIL WATER BALANCE

The estimation of K_s requires a daily water balance computation for the root zone. Schematically the root zone can be presented by means of a container in which the water content may fluctuate. To express the water content as root zone depletion is useful. It makes the adding and subtracting of losses and gains straightforward as the various parameters of the soil water budget are usually expressed in terms of water depth. Rainfall, irrigation and capillary rise of groundwater towards the root zone add water to the root zone and decrease the root zone depletion. Soil evaporation, crop transpiration and percolation losses remove water from the root zone and increase the depletion. The daily water balance, expressed in terms of depletion at the end of the day is:

$$D_{r,\,i} = D_{r,\,i-1} - (P - RO)_i - I_i - CR_i + ET_{c,\,i} + DP_i$$

where $D_{r,\,i}$ root zone depletion at the end of day i [mm],

$D_{r,\,i-1}$ water content in the root zone at the end of the previous day, $i - 1$ [mm],

P_i precipitation on day i [mm],

RO_i runoff from the soil surface on day i [mm],

I_i net irrigation depth on day i that infiltrates the soil [mm],

CR_i capillary rise from the groundwater table on day i [mm],

$ET_{c,\,i}$ crop evapotranspiration on day i [mm],

DP_i water loss out of the root zone by deep percolation on day i [mm].

Limits on $D_{r,\,i}$

Water can be stored in the root zone until field capacity is reached. Although following heavy rain or irrigation the water content might temporally exceed field capacity, the total amount of water above field capacity is assumed to be lost the same day by deep percolation, following any ET for that day. By assuming that the root zone is at field capacity following heavy rain or irrigation, the minimum value for the depletion $D_{r,i}$ is zero.

As a result of percolation and evapotranspiration, the water content in the root zone will gradually decrease and the root zone depletion will increase. In the absence of any wetting event, the water content will steadily reach its minimum value θ_{WP}. At that moment no water is left for evapotranspiration in the root zone, K_s becomes zero, and the root zone depletion has reached its maximum value TAW. The limits imposed on $D_{r,i}$ are consequently:

$$0 \le D_{r,i} \le TAW$$

Initial Depletion

To initiate the water balance for the root zone, the initial depletion $D_{r,i-1}$ should be estimated. The initial depletion can be derived from measured soil water content by:

$$D_{r,i-1} = 1000(\theta_{FC} - \theta_{i-1})Z_r$$

where θ_{i-1} is the average soil water content for the effective root zone. Following heavy rain or irrigation, the user can assume that the root zone is near field capacity, i.e., $D_{r,i-1} \approx 0$.

Precipitation (P), Runoff (RO) and Irrigation (I)

P_i is equivalent to daily precipitation. Daily precipitation in amounts less than about 0.2 ET_o is normally entirely evaporated and can usually be ignored in the water balance calculations especially when the single crop coefficient approach is being used. Ii is equivalent to the mean infiltrated irrigation depth expressed for the entire field surface. Runoff from the surface during precipitation can be predicted using standard procedures from hydrological texts.

Capillary Rise (CR)

The amount of water transported upwards by capillary rise from the water table to the root zone depends on the soil type, the depth of the water table and the wetness of the root zone. CR can normally be assumed to be zero when the water table is more than about 1 m below the bottom of the root zone. Some information on CR was presented in FAO Irrigation and Drainage Paper No. 24. CR will be a topic in a future FAO publication.

Evapotranspiration (ET_c)

Where the soil water depletion is smaller than RAW, the crop evapotranspiration equals $ET_c = K_c\ ET_o$. As soon as $D_{r,i}$ exceeds RAW, the crop evapotranspiration is reduced and ET_c can be computed from Equation.

Deep percolation (DP)

Following heavy rain or irrigation, the soil water content in the root zone might exceed field capacity. In this simple procedure it is assumed that the soil water content is at θ_{FC} within the same day of the wetting event, so that the depletion $D_{r,\,i}$ in Equation becomes zero. Therefore, following heavy rain or irrigation

$$DP_i = (P_i - RO_i) + I_i - ET_{c,\,i} - D_{r,\,i-1} \geq 0$$

As long as the soil water content in the root zone is below field capacity (i.e., $D_{r,\,i} > 0$), the soil will not drain and $DP_i = 0$.

The DP_i term in Equations is not to be confused with the $DP_{e,\,i}$ term used in Equations for the evaporation layer. Both terms can be calculated at the same time, but are independent of one another.

FORECASTING OR ALLOCATING IRRIGATIONS

Irrigation is required when rainfall is insufficient to compensate for the water lost by evapotranspiration. The primary objective of irrigation is to apply water at the right period and in the right amount. By calculating the soil water balance of the root zone on a daily basis the timing and the depth of future irrigations can be planned. To avoid crop water stress, irrigations should be applied before or at the moment when the readily available soil water is depleted ($D_{r,\,i} \leq$ RAW). To avoid deep percolation losses that may leach relevant nutrients out of the root zone, the net irrigation depth should be smaller than or equal to the root zone depletion ($I_i \leq D_{r,\,i}$).

EFFECTS OF SOIL SALINITY

Salts in the soil water solution can reduce evapotranspiration by making soil water less "available" for plant root extraction. Salts have an affinity for water and hence additional force is required for the crop to extract water from a saline soil. The presence of salts in the soil water solution reduces the total potential energy of the soil water solution. In addition, some salts cause toxic effects in plants and can reduce plant metabolism and growth. A function is presented here that predicts the reduction in evapotranspiration caused by salinity of soil water. The function is derived by combining yield-salinity equations from the FAO Irrigation and Drainage Paper N°29 with yield-ET equations from FAO Irrigation and Drainage Paper N°33. The resulting equation provides a first approximation of the reduction in evapotranspiration expected under various salinity conditions.

There is evidence that crop yield and transpiration are not as sensitive to low osmotic potential as they are to low matric potential. Under saline conditions, many plants are able to partially compensate for low osmotic potential of the soil water by building up higher internal solute contents. This is done by absorbing ions from the soil solution and by synthesizing organic osmolytes. Both of these reactions reduce the impact of osmotic potential on water availability. However, synthesis of organic osmolytes does require expenditure of metabolic energy. Therefore plant growth is often reduced under saline conditions. The reduced plant growth impacts transpiration by reducing ground cover and is sometimes additionally due to partial stomatal closure.

Other impacts of salts in the soil include direct sodium and chloride toxicities and induced nutrient deficiencies. These deficiencies reduce plant growth by reducing the rate of leaf elongation, the enlargement, and the division of cells in leaves. The modality depends on the method of irrigation. With sprinkler irrigation, adsorption of sodium and chloride through the leaf can result in toxic conditions for all crop species. With surface or trickle irrigation, direct toxic conditions generally occur only in vine and tree crops; however, high levels of sodium can induce calcium deficiencies for all crop species.

Since salt concentration changes as the soil water content changes, soil salinity is normally measured and expressed on the basis of the electrical conductivity of the saturation extract of the soil (EC_e). The EC_e is defined as the electrical conductivity of the soil water solution after the addition of a sufficient quantity of distilled water to bring the soil water content to saturation.

EC_e is typically expressed in deciSiemens per meter ($dS\ m^{-1}$). Under optimum management conditions, crop yields remain at potential levels until a specific, threshold electrical conductivity of the saturation soil water extract ($EC_{e\ threshold}$) is reached. If the average EC_e of the root zone increases above this critical threshold value, the yield is presumed to begin to decrease linearly in proportion to the increase in salinity. The rate of decrease in yield with increase in salinity is usually expressed as a slope, *b*, having units of % reduction in yield per dS/m increase in EC_e.

All plants do not respond to salinity in a similar manner; some crops can produce acceptable yields at much higher soil salinity levels than others. This is because some crops are better able to make the needed osmotic adjustments that enable them to extract more water from a saline soil, or they may be more tolerant of some of the toxic effects of salinity. Salt tolerance for many agricultural crops are provided in the FAO Irrigation and Drainage Papers No. 33 and 48. The $EC_{e\ threshold}$ and slope *b* from these sources. There is an 8 to 10-fold range in salt tolerance of agricultural crops. The effect of soil salinity on yield and crop evapotranspiration is hence crop specific. The $EC_{e\ threshold}$ and *b* parameters were determined primarily in research experiments using nearly steady-state irrigation where soil water contents were maintained at levels close to field capacity. However, under most types of irrigation scheduling for sprinkler and surface irrigation, the soil water content is typically depleted to well below field capacity, so that the EC of the soil water solution, EC_{SW}, increases prior to irrigation, even though the EC of the saturation extract does not change.

The increased salt concentration in the soil water solution reduces the osmotic potential of the soil water solution (it becomes more negative), so that the plant must expend more metabolic energy and may exert more mechanical force to absorb water. In addition, metabolic and toxic effects of salts on plants may become more pronounced as the soil dries and concentrations increase.

However, the variation in soil water content during an irrigation interval has not been found to strongly influence crop evapotranspiration. This is because of the rise of soil water content to levels that are above that experienced under steady state irrigation early in a long irrigation interval.

There is a similar, counteractive decrease in soil water content later in a long irrigation interval. In addition, the distribution of salts in the root zone under low frequency irrigation can reduce salinity impacts during the first portion of the irrigation interval. Also, under high frequency irrigation of the soil surface, soil evaporation losses are higher. Consequently, given the same application depth, the leaching fraction is reduced. For these reasons, the length of irrigation interval and the change in EC of soil water during the interval have usually not been found to be factors in the reduction of ET, given that the same depths of water are infiltrated into the root zone over time.

In some cases, increased evaporation under high frequency irrigation of the soil surface can counteract reductions in K_c caused by high EC_e of the root zone. Under these conditions, the total K_c and ET_c are not very different from the non-saline, standard conditions under less frequent irrigation, even though crop yields and crop transpiration are reduced. Because of this, under saline conditions, the K_s reducing factor should only be applied with the dual K_c approach.

YIELD-SALINITY RELATIONSHIP

A widely practiced approach for predicting the reduction in crop yield due to salinity has been described in the FAO Irrigation and Drainage Paper N°29. The approach presumes that, under optimum management conditions, crop yields remain at potential levels until a specific, threshold electrical conductivity of the soil water solution is reached. When salinity increases beyond this threshold, crop yields are presumed to decrease linearly in proportion to the increase in salinity. The soil water salinity is expressed as the electrical conductivity of the saturation extract, EC_e. In equation form, the procedure followed in FAO Irrigation and Drainage Paper N°29 is:

$$\frac{Y_a}{Y_m} = 1 - (EC_e - EC_{e\ \text{treshold}})\frac{b}{100}$$

for conditions where $EC_e > EC_{e\ \text{threshold}}$ where:

Y_a actual crop yield

Y_m maximum expected crop yield when $EC_e < EC_{e\ \text{threshold}}$

EC_e mean electrical conductivity of the saturation extract for the root zone [dS m^{-1}]

$EC_{e\ \text{threshold}}$ electrical conductivity of the saturation extract at the threshold of EC_e when crop yield first reduces below Y_m [dS m^{-1}]

b reduction in yield per increase in EC_e [%/(dS m^{-1})]

Values for $EC_{e\ \text{threshold}}$ and *b* have been provided in the FAO Irrigation and Drainage Paper N°29 and 48 and are for many agricultural crops.

Salinity-yield data from the FAO Irrigation and Drainage papers Nos. 29 and 48 were mostly from studies where soil water content was held at about-3 m potential (–30 kPa) or higher at the 0.3 to 0.6 m depth, depending on the crop. However, these papers indicate that for most crops, the data are transferable to typical field situations where the readily available soil water (RAW) is depleted between irrigations.

YIELD-MOISTURE STRESS RELATIONSHIP

A simple, linear crop-water production function was introduced in the FAO Irrigation and Drainage Paper N°33 to predict the reduction in crop yield when crop stress was caused by a shortage of soil water:

$$\left(1 - \frac{Y_a}{Y_m}\right) = K_Y\left(1 - \frac{ET_{c\ \text{adj}}}{ET_c}\right)$$

where: K_y a yield response factor [–]

$ET_{c\ \text{adj}}$ adjusted (actual) crop evapotranspiration [mm d^{-1}]

ET_c crop evapotranspiration for standard conditions (no water stress) [mm d^{-1}]

K_y is a factor that describes the reduction in relative yield according to the reduction in ET_c caused by soil water shortage. In FAO N°33, K_y values are crop specific and may vary over the growing season. In general, the decrease in yield due to water deficit during the vegetative and ripening period is relatively small, while during the flowering and yield formation periods it will be large. Values for K_y for individual growth periods and for the complete growing season have been included in the FAO Irrigation and Drainage Paper N°33.

COMBINED SALINITY-ET REDUCTION RELATIONSHIP

No Water Stress (D_r < RAW)

When salinity stress occurs without water stress, Equations can be combined and solved for an equivalent K_s, where $K_s = ET_{c\ adj}/ET_c$:

$$K_s = 1 - \frac{b}{K_Y 100}(EC_e - EC_{e\ treshold})$$

for conditions when $EC_e > EC_{e\ threshold}$ and soil water depletion is less than the readily available soil water depth (D_r < RAW). D_r and RAW.

WITH WATER STRESS (D_r > RAW)

When soil water stress occurs in addition to salinity stress, Equation are combined to yield:

$$K_s = \left(1 - \frac{b}{K_Y 100}(EC_e - EC_{e\ treshold})\right)\left(\frac{TAW - D_r}{TAW - RAW}\right)$$

for conditions when $EC_e > EC_{e\ threshold}$ and D_r > RAW. The impact of salinity reduction on K_s as salinity increases. Note that the approach presumes that RAW (and p) do not change with increasing salinity. This may or may not be a good assumption for some crops.

Limitations

Because the impact of salinity on plant growth and yield and on crop evapotranspiration is a time-integrated process, generally only the seasonal value for K_y is used to predict the reduction in evapotranspiration. There are K_y values in FAO Irrigation and Drainage paper N°33 for only about 23 crops. For many crops, the seasonal K_y is nearly 1. For crops where K_y is unknown, the user may use K_y = 1 in Equations or may select the K_y for a crop type that has similar behaviour.

Equations are suggested as only approximate estimates of salinity impacts on ET, and represent general effects of salinity on evapotranspiration as occurring over an extended period of time (as measured in weeks or months). These equations are not expected to be accurate for predicting ET_c for specific days. Nor do they include other complicating effects such as specific ion toxicity.

The equations presented may not be valid at high salinity, where the linear relationships between EC_e, crop yield and K_s may not hold. The use of Equations should usually be restricted to $EC_e < EC_{threshold} + 50/b$. In addition, the equations predict $Y_a = 0$ before $K_s = 0$ when $K_y > 1$ and vice versa.

As indicated earlier, reduction in ET_c in the presence of soil salinity is often partially caused by reduced plant size and fraction of ground cover. These effects are largely included in the coefficient values. Therefore, where plant growth is affected by salinity and Equations are applied, no other reductions in K_c are required, for example using LAI or fraction of ground cover

ET_c UNDER VARIOUS MANAGEMENT PRACTICES

EFFECTS OF SURFACE MULCHES

Mulches are frequently used in vegetable production to reduce evaporation losses from the soil surface, to accelerate crop development in cool climates by increasing soil temperature, to reduce erosion, or to assist in weed control. Mulches may be composed of organic plant materials or they may be synthetic mulches consisting of plastic sheets.

PLASTIC MULCHES

Plastic mulches generally consist of thin sheets of polyethylene or a similar material placed over the ground surface, especially along the plant rows. Holes are cut into the plastic film at plant spacings to allow the plant vegetation to emerge. Plastic mulches can be transparent, white or black. Colour influences albedo mainly during the early stages of the crop. However, as the intention is to use a simple procedure for adjusting K_c for mulched crops, no distinction is made between the different types of plastic mulches.

Table. Approximate reductions in K_c and surface evaporation and increases in transpiration for various horticultural crops under complete plastic mulch as compared with no mulch using trickle irrigation

Crop	*Reduction in K_c (%)*	*Reduction in Evaporation (%)*	*Increase in Transpiration (%)*
Squash	5-15	40-70	10-30
Cucumber	15-20	40-60	15-30
Cantaloupe	5-10	80	35
Watermelon	25-30	90	-10
Tomato	35	not reported	not reported
Average	10-30	50-80	10-30

Plastic mulches substantially reduce the evaporation of water from the soil surface, especially under trickle irrigation systems. Associated with the reduction in evaporation is a general increase in transpiration from vegetation caused by the transfer of both sensible and radiative heat from the surface of the plastic cover to adjacent vegetation.

Even though the transpiration rates under mulch may increase by an average of 10-30% over the season as compared to using no mulch, the K_c values decrease by an average of 10-30% due to the 50-80% reduction in soil evaporation. A summary of observed reductions in K_c, in evaporation, and increases in transpiration over growing seasons for five horticultural crops. Generally, crop growth rates and vegetable yields are increased by the use of plastic mulches.

Single Crop Coefficient, K_c

To consider the effects of plastic mulch on ET_c, the values for $K_{c\ mid}$ and $K_{c\ end}$ for the horticultural crops can be reduced by 10-30%, depending on the frequency of irrigation (use the higher value for frequent trickle irrigation). The value for $K_{c\ ini}$ under mulch is often as low as 0.10. When the plastic mulch does not entirely cover the soil wetted by the drip emitters, or where substantial rainfall occurs, then the reduction in $K_{c\ mid}$ or $K_{c\ end}$ will be less, in proportion to the fraction of wet surface covered by the mulch.

Dual Crop Coefficient, $K_{cb} + K_e$

When estimating basal K_{cb} for mulched crops, less adjustment is normally needed to the K_{cb} curve, being of the order of a 5-15% reduction in K_{cb}, as it is generally understood that the 'basal' evaporation of water from the soil surface is less with a plastic mulch, though the transpiration is increased. The effect on K_{cb} could be greater in some situations and with some types of low density crops. Local calibration of K_{cb} (and K_c) for use with mulch culture is encouraged.

When calculating the soil evaporation coefficient K_e with plastic mulch, the f_w should represent the relative equivalent fraction of the ground surface that can contribute to evaporation through the vent holes in the plastic cover and to the fraction of surface that is wetted, but is not covered by the mulch. The effective area of vent holes is normally two to four times the physical area of the vents (or even higher) to account for vapour transfer from under the sheet.

ORGANIC MULCHES

Organic mulches are often used with orchard production and with row crops under reduced tillage operations. Organic mulches may consist of unincorporated plant residues or foreign material imported to the field such as straw. The depth of the organic mulch and the fraction of the soil surface covered can vary widely. These two parameters will affect the amount of reduction in evaporation from the soil surface.

Single Crop Coefficient, K_c

A general rule when applying K_c is to reduce the amount of soil water evaporation by about 5% for each 10% of soil surface that is effectively covered by an organic mulch.

For example, if 50% of the soil surface were covered by an organic crop residue mulch, then the soil evaporation would be reduced by about 25%.

- In the case of $K_{c\ ini}$, which represents mostly evaporation from soil, one would reduce $K_{c\ ini}$ by about 25% in this situation.
- In the cases of $K_{c\ mid}$ and $K_{c\ end}$, one would reduce these values by 25% of the difference between ($K_{c\ mid}$ – $K_{cb\ mid}$) and ($K_{c\ end}$ – $K_{cb\ end}$). Generally, the differences between values are only 5-10% so that the adjustment to $K_{c\ mid}$ and $K_{c\ end}$ to account for an organic mulch may not be very large.

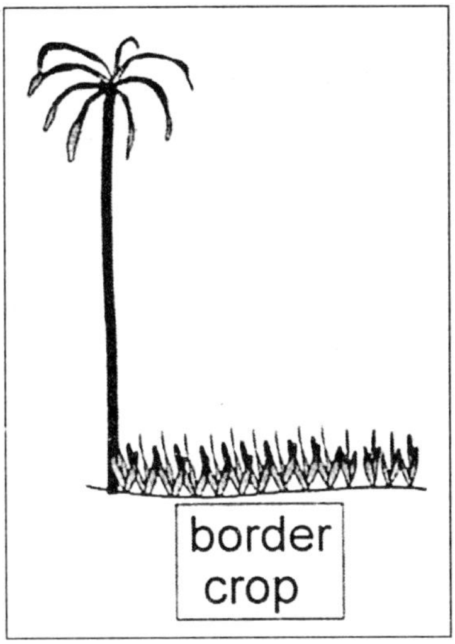

Fig. Different Situations of Intercropping

Dual Crop Coefficient, $K_{cb} + K_e$

When applying the approach with a separate water balance of the surface soil layer, the magnitude of the evaporation component (K_e ET_o) should be reduced by about 5% for each 10% of soil surface covered by the organic mulch. K_{cb} is not changed.

These recommendations are only approximate and attempt to account for the effects of partial reflection of solar radiation from residue, microadvection of heat from residue into the soil, lateral movement of soil water from below residue to exposed soil, and the insulating effect of the organic cover. As these parameters can vary widely, local observations and measurements are required if precise estimates are required.

INTERCROPPING

Intercropping refers to the situation where two different crops are grown together within one field. For the estimation of the crop coefficient, a distinction is made between:

- Contiguous vegetation, where the canopies of the two crops intermingle at some height (e.g., corn and beans intercropping).
- Overlapping crops, where the canopy of one crop is well above that of the other so that the canopies cannot be considered to be contiguous (e.g., date trees overlapping pomegranate trees at an oasis).
- Border crops, where tall crops such as windbreaks border fields of shorter crops, or high trees border a field crop.

There is an upper limit to the energy available to evaporate water.

This is represented by $K_{c\ max}$ for all crops in cultivated fields larger than 3-5 ha:

$$K_{c\max} = \max\left(\left\{1.2 + [0.04(u_2 - 2) - 0.004(RH_{\min} - 45)]\left(\frac{h}{3}\right)^{0.3}\right\}.\{K_{cb} + 0.05\}\right)$$

where h is the height for the taller crop. Under all conditions when combining crop coefficients for multiple crops, K_c should be constrained by this upper bound ($K_c \leq K_{c\ max}$).

CONTIGUOUS VEGETATION

Where the taller crop has canopy foliage that extends down to the same elevation as that of the top of the shorter crop, the vegetation canopy can be considered to be contiguous. For example, in Africa and South America, maize and beans are frequently intercropped as contiguous vegetation, with one row of maize planted per one or more rows of beans. Another example is the cultivation of five to seven rows of wheat intercropped with three rows of maize in many areas of China.

Similar Ground Cover

Where the leaf area or fraction of ground covered by the vegetation (f_c) is similar for each crop, the K_c for the taller crop (if this K_c is higher) can be taken to represent the entire field. The taller crop will act in some sense as a clothesline so that K_c (and ET_c) for the taller crop per unit of ground area.

However, the K_c (and ET_c) for the shorter crop will be reduced due to the windbreak effect by the taller crop. As a result, the K_c for the field as a whole may be similar to the weighted average of the K_c values for the two crops, the total K_c may more closely follow the K_c predicted for a field sown entirely to the taller crop ($K_{c\ field} \approx K_{c\ taller\ crop}$). Yields for the shorter crop may be reduced relative to those for single cultivar production due to the effects of shading by the taller crop and the competition for soil water.

Different Ground Cover

Where the fractions of ground covered by each crop are different, the K_c for an intercropped field can be estimated by weighting the K_c values for the individual crops according to the fraction of area covered by each crop and by the height of the crop:

$$K_{c\ field} = \frac{f_1 h_1 K_{c1} + f_2 h_2 K_{c2}}{f_1 h_1 + f_2 h_2}$$

where f_1 and f_2 are the fractions of the field surface planted to crops 1 and 2, h_1 and h_2 are the heights of crops 1 and 2, and K_{c1} and K_{c2} are the K_c values for crops 1 and 2.

OVERLAPPING VEGETATION

Where intercropping entails overlapping of spacings, the canopy of one crop is well above the other. This is the case, for example in southern California, where citrus trees are planted in date palm groves. Where a normal dense spacing is used for both the dates and for the citrus trees, the K_c may increase as the density of the combined vegetation increases, proportional to the increase in LAI with maximum K_c constrained by either $K_{c\ max}$ or by $K_{cb\ full}$ unless the total field area is small so that there is an additional clothesline or oasis effect.

CLOTHESLINE AND OASIS EFFECTS

Under the clothesline effect, where vegetation height is greater than that of the surroundings (different roughness conditions), or under the oasis effect, where vegetation has higher soil water availability than the surroundings (different moisture conditions), the peak K_c values may exceed the 1.20-1.40 limit. The user should exercise caution when extrapolating ET measurements taken from these sorts of vegetation stands or plots to larger stands or regions as an overestimation of regional ET may occur.

Small expanses of tall vegetation that are surrounded by shorter cover can exhibit a clothesline effect. This occurs where turbulent transport of sensible heat into the canopy and transport of vapour away from the canopy is increased by the 'broadsiding' of wind horizontally into the taller vegetation. In addition, the internal boundary layer above the vegetation may not be in equilibrium with the new surface. Therefore, ET from the isolated expanses, on a per unit area basis, may be significantly greater than the corresponding ET_o computed for the grass reference. Examples of the clothesline or oasis effects would be ET from a single row of trees surrounded by short vegetation or surrounded by a dry non-cropped field, or ET from a narrow strip of cattails (a hydrophytic vegetation) along a stream channel. K_c values up to and exceeding two have been recorded for such situations.

Where ET estimates are needed for such small, isolated expanses of vegetation surrounded by shorter cover (clothesline effect) or dry land (oasis effect), then the K_c may exceed the grass reference by 100% or more. Estimates of K_c for the expanses of vegetation should contain u_2, RH_{min} and h parameters to adjust K_c values, and parameters expressing the aridity of the surrounding area, the general width of the vegetation stand and the ability of the wind to penetrate into the vegetation. The equation should also consider the LAI of the vegetation to account for the ability of the vegetation to conduct and transpire the amount of water demanded by the clothesline/climatic condition. An upper limit of 2.5 is usually placed on K_c to represent an upper limit on the stomatal capacity of the vegetation to supply water vapour to the air stream under the clothesline or oasis conditions. For vegetation with a great leaf resistance, such as for some types of desert vegetation or trees, the upper limit should be multiplied by a resistance correction factor, F_r.

Curves of K_c for small areas of vegetation versus vegetation stand width, for conditions where u_2 = 2 m/s, RH_{min} = 30%, vegetation height = 2 m, and LAI = 3. The upper curve represents conditions where the specific vegetation is surrounded by dead vegetation, dry bare soil, or even gravel or asphalt.

In this situation, large amounts of sensible heat are generated from the surrounding area due to the lack of ET. Some of this sensible heat is advected into the vegetation downwind. The lower curve represents conditions where the vegetation is surrounded by well-watered grass. In this situation, there is much less sensible heat available from the surrounding area to increase ET (and K_c) of vegetation downwind. The influence of the aridity of the surroundings on the K_c for a small expanse is apparent. The two curves shown will change with changes in u_2, RH_{min}, h, and LAI.

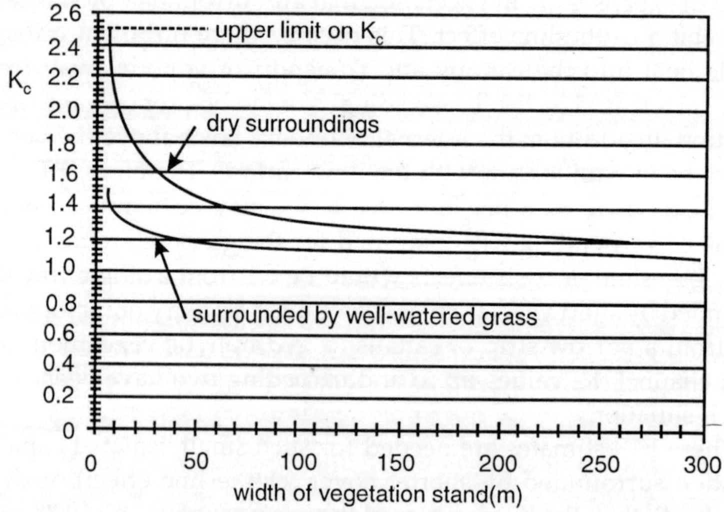

Fig. K_c curves for small areas of vegetation under the oasis effect as a function of the width of the expanse of vegetation for conditions where RH_{min} = 30%, u_2 = 2 m/s, vegetation height (h) = 2 m and LAI = 3

ET estimates from large expanses of vegetation or from small expanses of vegetation that are surrounded by mixtures of other vegetation having similar roughness and moisture conditions should almost always be less than or equal to 1.4 ET_o, even under arid conditions.

For tall wind breaks, such as single rows of trees, an approximate estimate for K_c is:

$$K_c = \min\left(1.2 + \frac{f_r h_{canopy}}{Width}, 2.5\right)$$

where F_r stomatal resistance correction factor

h_{canopy} mean vertical height of canopy area [m]

Width width (horizontal thickness) of the windbreak [m]

The K_c = 2.5 limit imposed in Equation represents an approximate upper limit on ET_c of trees per unit ground area. However, this value has large uncertainty. Because availability of soil water may limit evapotranspiration from wind breaks, a soil water balance and calculation of the K_s stress factor should be conducted.

MANAGEMENT INDUCED ENVIRONMENTAL STRESS

Many agricultural crops are intentionally water stressed during specific crop growth periods to encourage particular crop characteristics. The water stress is initiated by withholding or by reducing irrigations. In situations where this type of cultural management is practised, the K_c should be reduced to account for the reduction in evapotranspiration.

Environmental stress from soil water shortage, low soil fertility, or soil salinity can cause some types of plants to accelerate their reproductive cycle. In these situations, the length of the growing season may be shortened, particularly the mid-season period. Stress during the development period may increase the length of that period. Therefore, the length of the mid-season, L_{mid}, and perhaps the lengths of the development and late seasons may need to be adjusted for environmentally stressed or damaged vegetation.

Local research and observation is critical to identify the magnitudes and extent of these adjustments. Some examples of modifications to K_c and to lengths of growing periods are described below.

Alfalfa Seed

Some forage crops such as alfalfa that are grown for seed production are intentionally water stressed to reduce the amount of vegetation and to encourage increased production of flowers and seed. In areas subject to freezing winters, the reduction in K_c for deep rooted crops such as alfalfa depends upon the amount of water made available from precipitation during the dormant (winter) season and upon the amount of rainfall and limited irrigation during the growing season. Therefore, the effects of the intentional stress on the values for K_c should be modelled using the basal crop procedure presented and the K_s coefficient and water balance.

Cotton

In cotton production, soil water stress may be initiated during the development period to delay flower development and to encourage boll development. This practice retards the growth rate of the cotton plant and delays the date of full cover. For cotton, the attainment of full cover and the beginning of the mid-season generally occurs when the LAI reaches approximately three. When soil water stress and growth retardation is practised, full cover may occur after the beginning of flowering. The effect of stress during the development period on ET_c can be incorporated into the K_c curve by extending the length of the development period into the mid-season period. The length of the total season generally remains the same.

Sugar Beets

Sugar beets are frequently managed to initiate mild soil water stress during the late season period to dehydrate roots and concentrate sugars. A terminal irrigation may be needed just prior to harvest to assist in root extraction. When this type of water stress is practised, the value for $K_{c\ end}$ is reduced from 1.0 to 0.6.

Coffee

Coffee plants are often intentionally water stressed to reduce vegetation growth and to encourage development of berries. Under these conditions, K_c values should be reduced. In addition, coffee fields may be bordered by trees that serve as windbreaks. The effect of windbreaks is to reduce the K_c of the coffee plants due to a reduction in wind and solar radiation over the plants. The reduction in K_c could be significant where windbreaks are tall and frequent. However, the K_c for the entire field area, including the windbreaks, may be increased by the presence of the trees, relative to the values for K_c for coffee due to increased total leaf area of the coffee-tree combination and the increased aerodynamic roughness.

Tea

Initiation and development of new leaves on tea plants often occurs following the start of the rainy season. During the dry season, initiation of new leaves is slow or non-existent. The transpiration from older leaves is lower than for new leaves due to effects of leaf age on stomatal conductance. Therefore, the K_c, when leaves have aged (more than 2-3 months old), will be perhaps 10-20% lower. Similar to coffee, tea fields may be bordered by trees that serve as windbreaks. The effect of windbreaks is to reduce the K_c of the tea plants, but to potentially increase the K_c for the entire plantation, as described for coffee.

Olives

Growers may increase spacings of olive trees under rainfed conditions in areas with less rainfall. This is done to increase the ground area per tree that contributes infiltrated rain to transpiration of the tree. For example, in Tunisia, the spacing of olive trees changes from the north to the south, in proportion to annual rainfall. The tree spacing influences the K_c for the crop.

ET_c DURING NON-GROWING PERIODS

TYPES OF SURFACE CONDITIONS

The type and condition of the ground surface during non-growing periods dictates the range for ET_c. Where the surface is bare soil, then the K_c will be quite similar to the $K_{c\ ini}$ predicted. Where the surface is covered by nearly dead vegetation or some type of organic mulch or crop residue, then the K_c will be similar to that for agriculture that uses a surface mulch. Where the surface is covered by weed growth or growth of 'volunteer' plants, then the K_c will vary according to the leaf area or fraction of ground covered by the vegetation and by the availability of soil water. Where the surface is snow covered or frozen, then the K_c is difficult to predict and a constant value for ET_c may have to be assumed.

BARE SOIL

Single Crop Coefficient

Where the ground is left mostly bare following harvest, then the K_c following harvest will be strongly influenced by the frequency and amount of precipitation. K_c for bare soil can be calculated as $K_c = K_{c\ ini}$ where $K_{c\ ini}$ is calculated using the procedure.

Dual Crop Coefficient

Where a daily soil water balance can be applied, the user may elect to apply the dual K_c approach. In this situation, the topsoil layer may dry to very low water contents during periods having no precipitation. Therefore, the values for K_{cb} and for $K_{c\ min}$ in Equations should be set equal to zero. This provides for the opportunity to predict $ET_c = 0$ during long periods having no rainfall. This is necessary to preserve the water balance of the evaporation layer and of the root zone in total. The daily water balance calculation, given $K_{cb} = 0$, will provide the most accurate estimates of ET_c during the non-growing periods.

SURFACE COVERED WITH DEAD VEGETATION

Single Crop Coefficient

Where the ground surface has a plant residue or other dead organic mulch cover, or where part of the unharvested crop remains suspended above the surface in a dead or senesced condition, then the surface will respond similarly to a surface covered by mulch. In this case, K_c can be set equal to $K_{c\ ini}$ as predicted but the value for $K_{c\ ini}$ can be reduced by about 5% for each 10% of soil surface that is effectively covered by an organic mulch.

Dual Crop Coefficient

Evaporation from dead, wet vegetation can be substantial for a few days following a precipitation event. Therefore, in the dual K_c approach, the value for f_c should be set equal to zero to reflect the lack of green cover and f_w should be set equal to 1.0 to reflect the wetting of both soil and mulch cover by precipitation.

The dead mulch or vegetation will dry more quickly than would the underlying soil if it were exposed. In addition, the soil will be protected somewhat from evaporation by the dead mulch or vegetation cover. Therefore, total evaporation losses will be less than the TEW predicted from Equation. This can be accounted for by reducing the value for TEW by 5% for each 10% of soil surface that is effectively covered by an organic mulch. The value for REW should be limited to less than or equal to that for TEW.

SURFACE COVERED WITH LIVE VEGETATION

During frost-free periods following harvest, weeds may begin to germinate and grow. This vegetation is supplied with water from storage in the soil profile and from any rainfall. In addition, crop seed lost during harvest may germinate following rainfall events and add to the ground cover. The amount of ground surface covered by vegetation will depend on the severity of weed infestation; the density of the volunteer crop; tile frequency and extent of soil tillage; the availability of soil water or rain, and any damage by frost. The value for K_{cb} during the non-growing period can be predicted over time according to the amount of vegetation covering the surface. This can be done through estimates of LAI using Equation or estimates of the fraction of ground cover, f_c.

Single Crop Coefficient

In the single crop coefficient approach, the value for K_{cb} determined using procedures can be converted into an equivalent K_c by adding 0.05 to 0.15 according to the frequency of rainfall and surface wetting.

It is important that the K_c for vegetation during the non-growing period be limited according to die amount of soil water available to supply evapotranspiration. Otherwise, the law of conservation of mass will be violated. Under all conditions, the integration of K_c ET_o over the course of the non-growing period cannot exceed the sum of the precipitation occurring during the period plus any residual soil water in the root zone following harvest that can be depleted by the subsequent vegetation.

The root zone in this case is the root zone for the weed or volunteer crops. A daily soil water balance may provide for the best estimate of soil water induced stress and associated reduction in K_c and ET_c.

Dual Crop Coefficient

Under the dual crop coefficient approach, K_{cb} can be predicted according to the amount of surface that is covered by vegetation using Equation. Then, a full daily soil water balance of the topsoil together with a full daily soil water balance of the root zone can be employed. The soil water balances will automatically adhere to the law of conservation of mass, so that total ET_c from the weed or volunteer vegetation will not be overestimated. Again, because the topsoil layer may dry to below wilting point under conditions of sparse rainfall, the values for K_{cb} and $K_{c\ min}$ used in Equations should be set equal to zero. In this manner, the daily soil water balance with dual K_c calculations can progress throughout the non-growing period with good results.

FROZEN OR SNOW-COVERED SURFACES

Where the ground surface is snow covered or frozen, any vegetation will be largely non-responsive and non-contributing to ET_c, and the amount of ET_c will be closely related to the availability of free water at the surface and to the albedo of the surface.

The albedo of snow covered surfaces can range from 0.40 for old, dirty snow cover to 0.90 for fresh, dry snow. Therefore, the ET_c for snow cover will be less than ET_o for grass, as 25-85% less shortwave energy is available. In addition, some energy must be used to melt the snow before evaporation.

The use of ET_o under such conditions is of limited value, as the assumption of conditions sustaining a green grass cover is violated. It is even possible to obtain negative values for ET_o on some winter days where the longwave radiation from the surface is large and the vapour pressure deficit is small. It is under these conditions that net condensation of water from the atmosphere is possible. This would be similar to negative evaporation.

Given the limited value of ET_o (or even ET_p) under snow covered or frozen conditions, a single, average value may be best used to predict ET_c. The latitude of Kimberly is 42°N and the elevation is about 1200 m. Over the six-year study period, the ground was 50% covered by snow for 25% of the time from 1 October to 30 March. The ground, when exposed, was frozen about 50% of the time. The K_c averaged 0.25 during periods when the soil was not frozen but where frosts were occurring. When the ground had 50% snow cover or greater, the ET_c averaged only 0.4 mm/day. Wright found that over the six-month non-growing period, total cumulative ET_c exceeded precipitation by about 50 mm.

The mean measurements of ET_c during the 1985-1991 study period. The measurements have high correspondence to the total shortwave radiation energy available on a clear day, R_{so}, estimated as 0.75 R_a. There is some lag between ET_c and R_{so} and R_s caused by cooler temperatures in January - March as compared to the October - December period. The ET_c/R_{so} ratio averaged only 0.17 over the six-month period, and averaged 0.11 from 1 Dec. - 10 Mar. The ET_c/R_s ratio averaged 0.23 over the six-month period, and averaged 0.15 from 1 Dec. - 10 Mar.

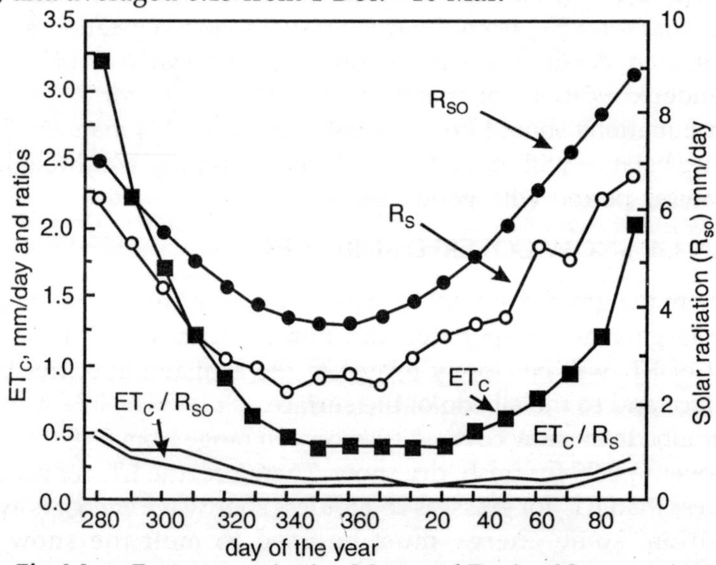

Fig. Mean Evapotranspiration Measured During Non-growing

A similar study conducted in Logan, Utah, the United States (latitude 41.6°N, elevation 1350 m) over an eight-year period showed that ET_c varied widely with soil surface wetness and air temperature during the winter months. The 'average' K_c from November to March was 0.5 for days having no snow cover. For days with snow cover, 'Et_c' ranged from 0 to 1.5 mm/day. Similarly, K_c is about 0.4 for winter wheat during frozen periods in the region of northern China (latitude near 39°N.

Single Crop Coefficient

The above procedure can provide estimates for the single K_c during non-growing season periods having snow cover or freezing conditions. However, the actual value for K_c is known to vary widely and will be less when water is less available at the soil surface.

Dual Crop Coefficient

A daily soil water balance using the dual crop coefficient approach is necessary to accurately predict ET_c under freezing and snow cover conditions. In the dual crop coefficient method, a daily water balance is conducted for the topsoil and the estimate for K_c can be reduced according to available water. However, in addition to the limited validity of the concept of ET_0 under frozen or snow covered conditions, the evaporation coefficient, K_e, may be reduced when the ground surface is frozen, as the water in a frozen state is less available.

Chapter 12

Institutional Economics and Irrigation Policy

The first objective of this chapter is to provide a summary of the book's institutional economic content–to recapitulate the broad lines of the text's economic reasoning. The second objective is to make proposals, not about irrigated agriculture in general but about the application of institutional economics to irrigation policy, including the design of policy instruments. The hydrosocial activities within this cycle are the appropriation of water for irrigation, its storage, its distribution to and from storage, its use in farmers' fields and the drainage of the water not consumed in evapotranspiration.

The text's economic paradigm is institutional economics, also known as political economy, rather than neoclassical economics or Marxist economics. In terms of method, a theoempiric approach is taken, a cyclical movement between reflection and fieldwork, fieldwork and reflection. In addition, the philosophical distance from the natural sciences is stressed–in studying human society we explore the actions of purposive, reflexive agents and the families and institutions these individuals constitute.

We may pursue useful theories but never universal laws. Irrigation organizations as well as those that support them in various ways, such as MFIs and research and policy bodies, should appraise the potential of institutional economics in the policy area of water for agriculture. If their appraisal is positive, they should draw on this paradigm in their work.

THE DEMAND FOR IRRIGATION SERVICES

Farmers are enormously varied, of course, one from another. The simplest and most powerful distinction between them in any single cultivated region is likely to be between farming families and agribusinesses. In output terms farmers who market their production seek crop tonnage, product quality and appropriate crop composition and timing of the harvest for the market. Where soil water availability sourced by precipitation to the rootzone of plants is insufficient in quantity or timing for crops' needs between planting and harvesting, then irrigation is required. The supply and use of irrigation water in agriculture demands that the economist consider the efficiency with which this natural resource is appropriated, delivered and applied. Irrigation water efficiencies can all be written as output-input ratios and the first to be explored in this book is E_1, relative efficiency.

Here the input is measured as actual water supplied divided by the physical optimum for plant growth and output is measured as actual crop tonnage divided by the physical maximum, with all other factors such as seed and fertilizer optimized. Relative efficiency is an agronomic concept and a tricky one at that because it is the ratio of two ratios. It can be used to show, for a wide variety of crops, that a proportionate increase in water supply is matched by a similar proportionate increase in crop tonnage but that this response fades at higher levels of water supply. By implication there may be little agronomic or economic justification in seeking maximum plant yields.

The second irrigation water efficiency is E_2, requirements efficiency, the ratio of net irrigation requirements divided by the base supply. Requirements efficiency is a distributional management concept. Assuming that a crop's net irrigation requirements are in fact met, it indicates the base supply actually used to resource those requirements. For example, at a value of 1.0 the whole of the base supply is taken up by crop evapotranspiration; at a value of 0.5 the base supply is twice the net irrigation requirements and one-half of the base supply flows off as drainage or is lost to non-beneficial evapotranspiration. This measure can be of great use in "crop per drop" discussions *provided* it is recognized that drainage water in one zone of an irrigation district is not necessarily lost to the district as a whole.

The third irrigation water efficiency considered is E_3, tonnage efficiency. This is a planning concept of limited interest, a ratio of crop output to the base supply of irrigation water. It is a close cousin of the farmer's concept of yield-output tonnage per hectare. Tonnage efficiency can be used in predicting the water resource requirements of the agricultural sector in a future year. In itself it is of little economic interest since it reflects neither output value nor input cost.

- The irrigation authority levies no charge for the supply of water. Here the farmer regards irrigation water only as an agronomic input, not as an economic good.
- Farmers themselves are responsible for the irrigation water supply and bear the costs both of installing the infrastructure as well as operation, maintenance and management.
- The irrigation authority charges the farmer a volumetric price.

- The irrigation authority levies a charge for water such that the total charge payable varies *indirectly* with the volume used. The *warabundi* rotation of turns with water allocated in proportion to land is the best-known indirect charge.
- An irrigation service fee is levied that bears no relation to the volume of water used.

The cost of water to the farmer of a type that resembles the pricing system for most market goods and services assumed by economists. Even the *warabundi* approach could not act as a price influencing the volume of water demand per hectare.

These reflections on charging systems and water prices lead on to consideration of the demand function for irrigation water. This represents the quantity of water that would be purchased in a given market at each of a number of different prices. Economists measure the responsiveness of volume demanded to price change by the price elasticity of demand. This elasticity is defined, for any given price-quantity combination, as the proportionate change in quantity divided by the proportionate change in price. The very important result obtained is that when price is low (and quantity purchased high) the value of the price elasticity of demand is low, of the order of "0.1. This implies that a 10 per cent increase in price would produce only a 1 per cent fall in quantity demanded.

The farm budget is considered next; it is a tabulation of a farm's annual flow of expenditure and income. On the expenditure side, prime costs plus overhead costs are equal to total costs. Prime costs are defined as expenditure on those inputs used up in the daily round of farm work, including any spending on irrigation water. Overhead costs of production are non-prime costs and are often assumed to be invariant with the level of crop output.

The farm's gross margin per hectare is the excess of total income ('turnover') over total prime costs, divided by farm size and is the source of income to cover the farm's overheads. Turnover less total costs equals the annual surplus of the farming family or agribusiness.

The farmer's demand for water is itself derived from the market's demand for the farmer's outputs. In terms of economic theory the farmer may estimate for irrigated crop output a rather sophisticated water efficiency measure: the increase in farm tonnage following a modest addition to the base supply of irrigation water, i.e., *marginal tonnage efficiency* (E_4). Fieldwork often suggests it has a zero value.

When the value is known either to farmers or to agricultural extension workers, it can be useful in moderating the demand for irrigation water.

Using their knowledge of how farm turnover would increase with the additional output and how the prime costs of water would rise, farmers would have a marginal efficiency measure *in economic terms* (E_5). E_5 transposes E_4 from physical to cash terms and can be called *marginal turnover efficiency*.

From this perspective, an increase in the price of water would probably be followed by a reduction in the demand for water. Similarly, other things being equal and for a known value of the net irrigation requirement, then the greater the gross margin of any single crop, the greater will be the quantity of irrigation water purchased for its cultivation at any given price of water. Conversely, when the farmer makes choices between the crops to be grown, the higher the price of water, the less likely it is that crops with low values of E_3, E_4 or E_5 will be selected. Which one of these three efficiencies is used by farmers doubtless varies; in economic terms, E_5 is the preferred criterion.

Economic theory is by no means always close to farmer behaviour in this respect. *First,* such rational action is most effective only in the cases of ownsupply cost and volumetric price. *Second,* it is feasible only where the farmer has control over the volume of water he can apply to the land. *Third,* the farmer may not know the response of crop output to marginal change in water use, adopting a satisficing rather than an optimizing pattern of behaviour. *Fourth,* if the price of water is low or if the proportion that water outlays make up in total prime costs is small, the marginal adjustment of this input may be seen as unimportant, so the elasticity of demand for irrigation water would be negligibly low. Here, in respect of prime costs, irrigation water is not an economic commodity to the farmer but is simply an agronomic necessity.

So in all catchments and for all families and agribusinesses, water is certainly seen as a vital natural resource in crop production, but is less often viewed as an economic input. In the latter case, the volume demanded will not be sensitive to modest price manipulations by a catchment agency.

This has at least six components. *First* comes internal and external reuse. In the irrigation cycle of external reuse appears as base flow 4 and internal reuse as base flow 5. Reuse engineering is best interpreted as a supply-side innovation. At the same time, it brings about a lower aggregate demand for water by the community as a whole than would happen in the absence of reuse and, in that sense, reuse is also treated as a form of demand management.

Second we have use technology. Water-use technologies that promise to achieve a higher level of requirements efficiency (E_2) are sprinkler, drip and trickle irrigation. In practice, without adequate management, the new technologies can be less efficient than simpler surface methods.

The *third and fourth* components of demand management are land-use planning and environmental education. The *fifth* component is water pricing, bringing us back to the demand for water in the strict economic sense of a price-quantity function. The metering of water use and setting a price per unit quantity help to underpin all the first four demand management forms.

Sixth and last, food import policies can play a very substantial part in management of the demand for water, particularly in semi-arid and arid countries. If a national government takes the decision to restrain the growth or cut-back the size of the agricultural sector and to place a growing reliance on imports to feed the domestic population, the demand for irrigation water will be less than in the no-policy-change situation. Introduced two more measures of efficiency.

E_6, gross margin efficiency, is the ratio for a farmer of his gross margin to the base supply of irrigation water. It failed to explain cropping patterns in the Curu Valley. E_7, turnover efficiency, appeared in the Keveral Farm case study and is equal to the gross sales income of irrigated produce per cubic metre of irrigation water. It can be estimated at the farm, scheme and irrigation district levels.

With respect to irrigation policy proposals, this is the place to address the issue of charges to farmers for the irrigation water with which they are supplied. Own-supply irrigators excepted, *the third proposal* is that when all is for the best in the best of all possible worlds irrigation water should be volumetrically priced so that total charge income covers the total costs of government's or a user association's infrastructure, operation, maintenance and management. The users of this natural resource, those who benefit from its appropriation, should pay for the cost of supply, with the added advantage that the waste of the resource is reduced if not avoided completely.

Few citizens live in the best of all possible worlds, as Voltaire recognized (1759). Full-cost pricing may be technically costly in requiring accurate measurement of the flow of irrigation water to each farmer. Full-cost pricing may also be politically infeasible, particularly where farmers consider themselves disadvantaged by other national policies (such as subsidized grain imports) or where the urban élite corruptly escalates infrastructure costs to its own advantage.

By way of illustration of this last point it is worthwhile pointing out that Clay Wescott, a senior Asian Development Bank (ADB) public administration specialist, said in 2001 that in the ADB region public expenditure was rent-skimmed on a large scale. In governance assessments already completed in Thailand and Vietnam the ADB had found that approximately 30 per cent of public expenditure in both countries "vanished via fraud and corruption".

In this case charges should be set at least to cover OM&M costs and paid directly to the irrigation supply authority, not to the regional or national Department of Finance. If volumetric pricing is too ambitious for whatever reason, indirect fees should be explored based on irrigated area or crop type or time allocation. OM&M charges based on an indirect fee method is *the fourth proposal,* a fall-back solution where full-cost pricing is not feasible.

THE SUPPLY OF IRRIGATION SERVICES: THE INFRASTRUCTURE

The institutional economics of irrigation services and farmland drainage is built on two pillars: the demand for those services and their supply. On the supply side it is convenient to distinguish between the supply of irrigation and drainage infrastructures and the daily activities of operation, maintenance and management.

The analysis of the infrastructure and begins with the vital distinction between capital and current spending. *Capital expenditure* is defined as that where the resource purchased has an expected life of more than 12 months. This includes land; civil engineering infrastructures such as boreholes, reservoirs, works road networks, pumps and pipes; buildings and plant of all kinds; and vehicles. *Current expenditure* refers to purchased resources that are either immediately used up in the process of production or which have a life of 12 months or less. These include materials, chemicals, spare parts, electric power and labour time.

In exploring supply-side efficiency we need a second and alternative way of classifying costs because we need *to add together* measures of infrastructural and operational costs to arrive at a total annual cost figure. (capital to current cost would give nonsense results as these two refer to different time-frames; also their addition can lead to double-counting errors.) This second and alternative categorization is between prime and overhead costs.

Prime costs are defined as those used up in the daily supply of irrigation services and consist of the salaries and wages of the workforce, the costs of power, materials, spare parts and other consumables such as bought-in specialist inputs. *Overhead* costs of production are non-prime costs and are often assumed to be invariant with the level of supply. In the simplest case, where capital expenditure on land, infrastructure and plant has been funded 100 per cent by loans, this element of annual overhead costs would be set equal to the annual interest payable and the principal repayable on the debt incurred. *Total* costs per year are equal to prime plus overhead costs.

The most important concepts in the irrigation economist's tool-kit to explore supply-side efficiency are *cost functions*. These quantitative relationships describe the cost of supplying output in any time-period at each scale of output from 0 units up to the system's theoretical capacity. The ability to shift from one delivery level of irrigation water up to a higher level clearly depends on the time available to make the change. In *the short term,* increased daily output is possible through operational changes or by organizational innovations demanding new procedures, both of which are relatively straightforward in their introduction. In *the long term,* additional infrastructure is required either in new projects or for the expansion of existing capacity, with associated financial, resource and organizational planning.

For each level of output, one can calculate overhead cost per unit of output. In the short term average overhead costs fall (at a decelerating rate) as the unchanged level of cost is divided by ever-greater levels of production. So in the short term, supply-side efficiency in terms of the overhead costs of irrigation water supply, i.e., the supply costs of the head works and networks infrastructure per unit of water delivered, is achieved by operating the irrigation system at its full capacity.

The cost function in the long run has a more complex meaning. It represents what average overhead cost is likely to be for each successively higher level of output where each such level would be achieved by an addition to capacity through investment in the capital infrastructure. Such planned additions to the baseline capacity are mutually exclusive alternatives, not evolutionary developments.

Whereas short-term economies of capacity utilization are virtually universal, the existence of longterm economies of scale is by no means certain. Where they *do* exist, economists explain them with the concept *of indivisibility*. When pressed to define the meaning of indivisibility, economists say it is what gives rise to long-run scale economies. In our world of irrigation and drainage, when such economies are verified they are best understood in the language of civil and mechanical engineering, hydrology and hydraulics.

To grasp the meaning of scale economies in the long run, an evolutionary approach to time was ruled out above by assuming mutually exclusive future states. But, of course, the capital stock of an irrigation district *does* evolve through time (although not in the sense of the evolutionary scientist). Here we have an interest in successive rounds of investment and their interdependence. In terms of longterm planning we must weigh up the advantages of small-scale efficiencies in capacity utilization against the possible loss of scale economies *and* the possibility that the first round of investment may drive up the unit overhead costs of the second round. The *lumpiness* of water infrastructures is what makes a 10 m dam built now impose a higher construction cost for an additional

10 m capacity a year later–a time-series comparison. Lumpiness particularly characterizes the irrigation networks of pipes and canals.

As we have seen, the infrastructural supply of irrigation services requires a wide variety of real capital resources. So the actors responsible for service supply must find the capital finance for these resources' capital costs. John Briscoe has pointed out how neglected is the subject of irrigation's capital financing.

Historically the actors responsible for service supply have fallen into three groups: farming families financing their own infrastructural supply, agribusinesses in a similar position and public-sector irrigation and drainage authorities. Broadly speaking, the first of these has the smallest capital needs per actor, the third the largest and agribusinesses lie in the middle.

The variety of ways in which capital financing is accessed is great, both within and between these three groups. The same applies to the terms under which finance is made available, such as the maximum loan available, the number of years over which it must be repaid, the rate of interest charged and the security demanded in case of failure to meet the agreed repayments schedule. These terms are best understood in the light of each country's banking and credit system or, in the case of many public-sector irrigation and drainage agencies, with reference to the policies and practices of the international banks.

In particular, the World Bank's prodigious capacity to lend is matched by an ability to borrow of extraordinary magnitude. Most of the Bank's lending is long term; so too is the bulk of its borrowing. This takes place in the world market for bonds. The market rate of interest on bonds with a long-term redemption date has many determinants, of which one is the credit rating of the institution issuing the bond. Credit rating is provided by two major agencies and both give World Bank debt their highest possible rating. As a result the market is willing to accept a rate of interest from the Bank on newly issued capital denominated in US dollars only 100 or so basis points above that paid by the safest security in the world-bonds issued by the US Treasury.

With respect to irrigation (and other) projects, the institution that borrows from the Bank is typically national government, not the institution that is responsible for the construction and management of the new infrastructure. An intermediary stands between the Bank and the irrigation authority. As a result, if the project fails the Bank does not suffer financially. There is a break in the line of prudential responsibility for ensuring that the institution requiring loan finance for capital investment is capable of repaying its debt. For the primary lender this "break in the line" has important consequences:

- The risks associated with the Bank's staff underestimating a project's costs and overestimating benefits are virtually eliminated.
- A weakly-regulated space opens between the secondary lender (central or regional government) and the final borrower (the irrigation water supplier); corrupt transfers and secondary lender manipulation of the project are thereby facilitated.
- Project finance becomes available for the treasury functions of the secondary lender, exacerbating the fungibility dilemma.
- The credit terms between the primary lender and the secondary lender can differ significantly from those between the secondary lender and the irrigation water supplier.
- The institutional commitment of the primary lender to the project can atrophy once the loan is disbursed; a rentier mentality sets in. The discipline that usually accompanies bank lending is diminished.

This brings us to the fifth proposal on irrigation policy. The MFI headquarters (HQ) located in Washington, Manila, Luxembourg, etc., should each establish its own banking in those nations where it has major business. Each country-MFI would be responsible, with its MFI HQ, for processing and approving loan applications for all its infrastructural projects–including in our specific case applications from irrigation and farmland drainage supply organizations. With approved projects the finance sourced by the MFI HQ would be made available through the country-MFI on credit terms agreed with the infrastructural agency. A specified number of basis points would be added to the loan rate of interest in order to cover the full costs of operating the country-mfi.

The country-MFI would be responsible for all aspects of loan administration and would have defined procedures for monitoring projects not only in terms of their financial status but also their progress, their strengths and weaknesses, as civil engineering, hydrological and social institutions. The country-MFI would deal directly in this way only in the case of projects above a defined size in financial terms. Where smaller loans are financed, the country-MFI would handle these through a selected local financial intermediary known for its competence and honesty, such as a local bank.

OPERATION, MAINTENANCE AND MANAGEMENT

OM&M–the tasks of running the infrastructural supply headworks and networks between the locations at which irrigation water is socially appropriated through to the points at which the water reaches farmers' fields. As with infrastructural overheads it is likely that average short-term prime cost falls as the utilization of capacity rises. Total cost is the sum of overhead and prime cost so in the short run average total cost can be expected to be an exponential function, at its lowest at full-capacity operation.

This leads to an additional two definitions of productivity. E_8, field supply efficiency, is the average total cost per cubic metre of the volume of irrigation water that actually reaches farmers' fields after drainage and evaporation losses. E_9, marginal field supply efficiency, refers to the ratio of the addition to total costs as a result of new investment in irrigation infrastructure, divided by the addition to the field supply.

I define full-cost payment for irrigation water as being equal to E_8, whether the payment is made as own-supply cost or volumetric price or as an indirect fee or as a revenue-only charge. From the point of view of efficiency in water use, ownsupply cost and volumetric prices are to be preferred. Widespread evidence points to a considerable shortfall between recommended maintenance expenditure for public-sector irrigation schemes and the amounts actually spent. In the first years of a new scheme, maintenance neglect may have no dramatic effect.

But sooner or later maintenance underfunding begins to bite. The biggest losses come through crop output decline as the volume of irrigation water delivered eventually begins to decline. This may show itself as a fall in yields or in a diminution of the irrigated area. Moreover, farmers shift to lower value crops so as to be able to reduce risk by limiting the use of inputs. Impeded drainage can lead to waterlogging and salinity and an increase in the incidence of water-related diseases associated with blocked channels and stagnant water. This process brings with it the premature obsolescence of irrigation projects. The search for new capital funding begins in order to renovate the assets far earlier than would have been necessary under a satisfactory maintenance regime. A cycle of build-neglect-rebuild is in place.

Assigning responsibility for the neglect of maintenance is complex. Often farmers pay in water charges far less than the necessary working expenditure for the system. Or government may corruptly overstaff its irrigation department, where in any case maintenance is a relatively neglected child. It may also be the case that OM&M payments by farmers go straight to the Ministry of Finance rather than being recycled for their proper purpose. What becomes clear, particularly with government-owned surface-water irrigation schemes to have any positive impact a change in organizational culture is required.

This is the historical context that has made the literature of the management of common property resources seem of such great relevance. But to be effective this applied theoretical approach should include within its scope not only the users of the common resource but also the service suppliers. With reference to users we should recognize that, for any single actor, private interest may prevail over the common interest in respect of resource appropriation, resource maintenance and resource degradation. Work motivated by the private interests of the agent can stimulate long hours in the field, a sharp tactical appreciation of gains and losses from adaptive action and an eager search for innovative practices. In respect of the productivity of the common resource, these are powerful advantages. However, with the passage of time the common property resource itself and the patterns of productive activity on it may show signs of enfeeblement and impending long-term collapse. Now, the disadvantages imposed by each single agent's actions on other persons, families and institutions have become disproportionately large.

Such a trajectory may lead to demands for a reconstitution of the social relationships between actors in the management of the common property resource. Where government cannot or will not lead this collective activity, a non-governmental LECA may be created. The core objectives of the LEGA are likely to be fourfold.

First, the shared use of the resource should be recognized by the participant agents to be equitable. *Second,* the maintenance of the resource should be adequate to ensure its longterm viability. *Third,* the production of negative externalities should be sufficiently well regulated that their collective cost to the new institution's actors is acceptable and does not threaten systemic sustainability. *Fourth,* the transaction costs of meeting the first three objectives are acceptable to actors in the light of the benefits they bring.

There is a problem. A successful LECA brings all-round advantages to its actor-members. But every agent knows that if he and he alone continues to pursue his private interest, he reaps the benefits of unrestrained action as well as the benefits deriving from the collective agreement. This is known as the freerider problem. As freerider numbers grow, the advantages of collective action diminish and the LECA collapses.

So each LECA needs to engage in forms of moral persuasion in order that its actors honour the collective agreement made. It will also monitor members' activities to ensure that they do not breach the rules and, when this occurs, sanctions are exercised against such infractions. The collective also faces costs of bargaining, contract formulation and information search in pursuit of a collective economic strategy. Together, all these costs of motivation, control and co-ordination are termed transaction costs.

The last two analytic with the abstraction charges levied by government on individuals and institutions for the right to abstract surface water and groundwater. These payments are a form of economic rent. A taxonomy of charge-setting principles is set out: no charge, a revenue-maximizing charge, a market-clearing charge, an environmental regulation charge, a Pigovian charge and an incentives charge.

We argue that abstraction charges can contribute positively to three of the six fields of action for water resource planning in a sustainable society. These three are the protection of water's hydrocyclical capacity to renew its ground- and surface-water flows and stocks, the conservation of society's species and natural habitats in all their fresh and saltwater environments and the husbandry of water in its supply and use.

The sixth proposal of this is that governments that decide to introduce abstraction fees should do so in the form of full cost incentive charging. The annual income from abstraction charges should be hypothecated to the environmental regulator such that, when added to other income sources, abstraction fee income is sufficient to cover all the state's capital and current account expenditures on environmental regulation, research and database development, compensation payments, etc.

Where the full cost abstraction tariff still leaves an excess of demand for abstracted water over its licensed supply the charge should be raised so that market clearing takes place. Specification of the components of the abstraction charge should provide incentives for abstraction behaviour that is economically efficient and that avoids environmental degradation.

The case study of irrigation management transfer on the Alto Río Lerma in Mexico leads me to the seventh proposal for irrigation policy. Where a public-sector surface-water irrigation supply system is recognized to be failing, with little hope of a renaissance, government should give serious consideration to its replacement by an irrigation water users association. This should not be done without first establishing a viable procedure for the new association to access capital finance for its future infrastructural requirements. Furthermore, it should be recognized that such associations are known in some cases to succeed in their mission and in other cases to fail miserably. IWMI is the most well-informed source on the conditions for success.

The case study of tradeable abstraction rights (TARs) in Victoria, Australia, leads to the eighth proposal Governments should review the strength of the arguments for introducing tradeable abstraction rights at the catchment level. TARs have the potential to function as a politically sophisticated method for transferring water to higher productivity applications. But TAR legislation needs to move forward cautiously so as to avoid a number of potentially serious negative outcomes. Reallocation can have powerful environmental effects locally and at the basin level. Moreover, TARs between economic sectors, such as from agriculture to mining, can wipe out economic activity in the selling sector with destructive impacts on local communities. The market is a good servant but a bad master.

FARMLAND DRAINAGE SERVICES: DEMAND AND SUPPLY

However, by the stage of reaching the discussion of theory and fieldwork proposed to avoid the term in future as well as the scholasticism of discussion on its true nature. Instead we should simply recognize that drainage benefits a variety of catchment actors, the services commonly enjoyed in this way are non-exclusive and the measurement of the uptake of the service with its consequential advantages may be complex, open to dispute or just downright costly. So the payment system for drainage supply requires a design quite different from that of the bulk of goods and services in capitalist society in terms of both assessment and collection.

Farmers' demand for drainage services may arise because of a desire to avoid the waterlogging of their crops, to prevent soil salinization, to leach salts already present in the soil or to recharge a local aquifer. Once undertaken, drainage water may be collected and reused for irrigation purposes. Payments for drainage services, whether or not the farmland drained is also irrigated, have marked similarities with those made for irrigation services alone. So we can find instances of no cost, own-cost, volumetric fee, indirect fee or revenue-only charge. In irrigated areas it is commonly the case that the supply of irrigation services and the supply of drainage services are carried out by the same institution. Where this happens, both are likely to be financed by a single payment mechanism.

At the heart of drainage engineering we find a means whereby the natural, gravity-driven flows of rainfall and surplus irrigation water across and beneath farmland are substituted by man-made channels powered by gravity or pumps. As a result, from a civil engineering and hydraulics perspective, the supply infrastructure of drainage services bears a striking similarity to the abstraction and distributional infrastructures of irrigation supply.

The common features of the civil engineering of land drainage and that of irrigation water abstraction and distribution have important consequences in economic terms. Land drainage networks may require major capital expenditure and capital financing. We can also expect economies of scale because of the indivisibility of drainage infrastructure provision. Lumpiness is also likely to be a feature of drainage networks. The double punch of indivisibility and lumpiness means that medium-term and long-term planning for drainage offers considerable benefits.

Pumping stations need to be operated and their equipment maintained. Drains require continuous work to remove debris of all kinds, to dredge silt and to cut back plant growth in and on the edge of watercourses. Drainage water is an important medium for the diffusion of agricultural pollutants beyond the farming sector. The policy issues concern either the agricultural use of agrochemicals or the standards and technologies for the treatment of abstracted water destined for households and industry. Thus the subject area, however vital, essentially falls under the economics of agriculture and the economics of freshwater treatment, outside the field of the economics of irrigation and drainage.

The ninth proposal is that when government or an irrigation and drainage authority reviews its charge-setting system for either irrigation or drainage services, it consider NRV as an option. Net rateable value taxation is an indirect fee of great efficiency in collection because it is a charge on immovable and openly identifiable property. It can be applied for administrative areas of huge variation in size. It can be made completely transparent by the open publication of the valuation list in each area. In English law it falls directly on the user of property but there is no reason why it should not be implemented so that it is paid by the owner of farmland. Thereby it would be correlated with the landed wealth of the family or institution paying the charge. Critical necessities of such a tax are accurate valuation, periodic updating and an honest and competent profession of valuers.

Social Cost-benefit Analysis for Irrigation and Drainage Projects

The method concerns itself primarily with the costs and benefits of the activity from the point of view of the nation as a whole and its GDP. Over the full life of the project one identifies the incremental value of the flow of output in each year with the project in place in comparison with the situation without the project. The same with/without procedure is adopted for the flow of costs. Market prices are used for valuation purposes. These two net flows are set out on a spreadsheet. The difference between benefit and cost in each year is the net benefit flow. The market prices used should be constant (usually base year) prices rather than out-turn prices, in order to avoid the misleading effects on evaluation of general price inflation.

For each year during the project's gestation period and working life the net benefit flow is then recalculated in terms of discounted values. This is done through dividing each entry by $(1+d)^t$ where d is the social time rate of discount (STRD) and t is the project year (from 0 to n) in which the entry occurs. We now have a stream of NPV terms.

The justification for discounting falls not within the field of economics but of psychology. There seems to be an almost universal human propensity to regard a real benefit offered now as superior to the same real benefit offered later in time. Human life is beset with risks of injury, loss and death. That has always been the human condition and is doubtless the origin of a rather rational desire to receive now, when the probability of receipt is greatest, rather than later, when such probability is perceived to be less. But what should be the STRD chosen?. The definition of the half-life associated with a specific social time rate of discount needs careful specification. For any real benefit valued today at US$1 if available today, but valued today at only US$0.50 if forecast to be available in h years time, then h is the half-life.

The Ministry should set its STRD for the country's projects in the light of this impact and its knowledge of project risk at the country scale. In this way we may achieve a middle way between searching for sustainability in projects chosen whilst recognizing the real risks of project obsolescence and failure. A benchmark STRD of 5 per cent per year may be widely appropriate, which would give a half-life for net benefit entries of about 14 years.

In most cases of project evaluation project interdependence exists, where two (or more) projects are mutually exclusive for technical reasons or where a budget constraint imposes a limit on the total number of projects that can be financed. In these cases, for every proposal (including all mutually exclusive projects) we calculate the net benefit-investment ratio. This is simply done by dividing the full project life into two: the gestation period, when the NPV terms are consistently negative and the working period, when the NPV terms are, in most years, positive. The sum of the NPV terms for each period are separately calculated and that of the working period divided by that of the gestation period. The negative sign of the denominator is ignored.

All projects are then ranked from top to bottom, using the net benefit-investment ratio. Lower ranking mutually exclusive projects are eliminated and then all the remaining projects are approved off the top until the budget constraint blocks any further approvals. What we have done in the second case, that of financially interdependent choices, is to consider the key constraint as the budget funds available during the gestation period and then to rank all projects in terms of the productivity (in NPV terms) of these funds.

This maximizes the total NPV generated by the use of the funds available. The net benefit-investment ratio can also be considered as a means not of excluding projects but of ranking them in terms of their starting date. When applied to irrigation and drainage projects, the net benefit-investment ratio is a valuable output-input ratio and we refer to it as E_{10}, yet another measure of irrigation water efficiency.

The eleventh proposal is that evaluation units and consulting economists throughout the world should cease using the economic rate of return and the aggregate NPV in project choice, but should always use the NBIR. As is stated above, market prices are used to convert incremental real outputs and real costs into the value terms of the evaluation spreadsheet. Project evaluation should always do this for its first estimate of the NBIR. However, there may be cases where the market price of an input or output is not deemed to be the best measure available to assess the project's contribution to GDP.

This is usually where the market price of an input does not measure its opportunity cost adequately. Where a unit price other than the market price is used, so that the values of incremental outputs and costs are recalculated, we speak of shadow prices.

In the calculation of the NBIR, the objective is to identify real outputs and real costs and then to value these at market (or at shadow) prices. In the financial accounts of a project, where project cash-flows are the focus, certain transfer items usually appear. These include the capital sum of a loan, repayment of loan principal, payment of loan interest, receipt of government subsidy and payment of government tax. They represent only transfers of money between one actor and another, not the values of real outputs and inputs from the perspective of the national or regional economy's GDP. So, in estimation of the NBIR, these items would not be included.

Project evaluation is always exposed to estimation bias. In the assessment of project outcomes, specific irrigation actors–in the private or the public sphere-will always have a material interest in specific projects receiving approval and others being rejected. This is made clear with startling force in Reisner's (1993) history of water projects in the American West. Within the MFIs a quite specific form of estimation bias is known as the McNamara effect. When the supply of funds for projects is in full flood, pressures of an informal kind are placed on appraisers to inflate their calculation of the rate of return, so that more projects qualify for approval.

Project evaluation should be carried out in two units–an *ex ante* unit (EAU) and an *ex post* unit (EPU). The parent institution would make it clear that evaluation should avoid estimation bias. All projects would be evaluated on three occasions: by the EAU prior to loan approval/refusal, by the EPU at the end of the gestation period and again by the EPU some 3-7 years into the project's working life. Each evaluation would be signed off by the staff responsible and would be available for sale to the public. The EPU would report directly to the MFI's Board or, in the case of government, to the responsible Minister. The EPU would maintain a constantly updated record and review of the *ex ante* and *ex post* evaluation outcomes and what consistency exists between the two. There would be no transfer of staff between an EAU and an EPU. A monthly joint seminar would be chaired alternately by the heads of the EAU and the EPU on the methods and experience of project evaluation within the organization and elsewhere; this would be known as the "Unbearable Lightness of Being Forum".

- The evaluation of a defined capital project should always be based jointly on its economic, social and environmental effects in comparison with the no-project alternative.
- The fantasy of economists that the social and environmental impacts can be meaningfully evaluated in dollar terms should be dismissed.
- The appraisal should be honest.
- The evaluation should be based on values as well as on qualitative and quantitative information.
- At the country or region level, the decision to proceed (or not) with a project should be taken by a group, balanced in its interests and skills, that understands and represents the economic, social and environmental categories of benefit and cost. Engineers should be represented on that group since their expertise in the how of the project and technically feasible alternatives is invaluable
- The group should indicate in its conclusion the relative weight given to each of the three categories–economy, society and environment–in arriving at its decision as well as the principal means by which each category of impacts is evaluated.
- The information on which a decision is based should be comprehensive, intelligible and readily available to all citizens.

Water Allocation between Competing Uses

The hydrosocial balance and the hydrosocial change balance. The first of these sets out for any defined area, such as a region, the ways in which society appropriates water flows for outstream purposes and the ways in which these flows are used. It is shown that total net supply and total use are mathematically identical. This procedure gives us a baseline balance for a specified time-period, such as the year 2002.

In a similar manner one can produce the hydrosocial balance for a future, scenario year such as 2006. By subtracting the baseline balance from the scenario balance we have the hydrosocial change balance. This shows what changes between 2002 and 2006 will have to take place if the scenario balance is to be realized. The change balance is the planning tool par excellence.

The fourteenth proposal of this final is that water resource authorities should use the hydrosocial balance and the hydrosocial change balance as fundamental techniques of their planning process. The change balance is applicable to short-, medium- and long-term strategy and to comparison of typical years past and future. It clarifies the quantitative choices that a regional government or catchment agency will be forced to envisage, for example, in a future year of severe drought or as a result of climatic change. The change balances can provide comprehensive, synoptic, transparent alternative scenarios within a region in planning infrastructural investment, capital financing and demand management.

It is in this way that it handles the uncertainty that is a major issue in water resources planning, as well as side-stepping the inadequacy of project-by-project planning. To suppose that in a specific region, for a 5- or 10-year planning period, water resource managers developing alternative hydrosocial change balances become aware that there is a serious impending conflict. The potential conflict is between irrigation as a user of water on the one side and, on the other, the non-farming users as well as the instream requirements for lake and river flows-nature conservation, fishing, navigation and recreation. An era of allocation stress is fast approaching.

Conflict resolution over allocation stress may be possible through substantial additional investment in supply-side expansion, it may be feasible through demand management in the water-using sectors and it may be achieved through reuse and recycling strategies. But suppose the allocation of water to the irrigation sector is under review, with a view to reducing its absolute volume and its percentage of total use. In this case three strategic arguments may be made in defence of irrigation: its efficiency in the use of water, its contribution to sustainable rural livelihoods and its provision of food to a region.

The fifteenth proposal is that irrigation authorities should review these alternative measures, determine which of them are both useful and feasible to estimate and should deploy those selected in such authorities' planning activities. With respect to allocation conflicts between sectors, sector efficiency is a vital measure; it is defined as the value-added output of an economic sector divided by the base supply of water to that sector. There is a general presumption that sector efficiency is relatively low for irrigated agriculture, as is the ratio of employment to the base supply-"jobs per drop".

In brief, this analytic approach to change in rural areas can be said to be made up of three principal blocks of ideas. *First,* in their daily lives individuals draw on natural, social, human, physical and financial capital asset combinations. *Second,* society's organizations and processes powerfully influence people's access to capital assets and these assets' values. *Third,* in shaping their livelihood strategies to build a life and endure in the face of adverse trends and shocks, people's real options are fundamentally shaped by their asset status and the institutional framework. In the light of this approach, what can we say about a regional water resource policy that reduces significantly the volume of irrigation water supplied to the farming sector?

Clearly here we posit an adverse shock and a continuing negative trend for rural livelihoods. Farmers' access to the natural capital asset of the hydrological resource is cut back. At the same time, the values of assets such as farmland diminish. Farmer networks may be weakened as some suffer whilst others escape unscathed. The human capital embedded in the daily routines of agricultural practice is devalued. Physical capital becomes obsolescent overnight. Financial capital is eroded as income from the sale of produce disappears.

Our further proposal is that, where a national or regional water authority identifies potentially serious conflict over water allocation, government action must include one or all of the following measures:

- Review the possibility of expanding the water supply infrastructure, at the same time as taking fully into account the financial and environmental consequences of such projects.
- Push through management structures and practices that reduce basin supply losses and raise water efficiency in use in all sectors-agriculture, households, industry and urban services.
- In particular, pursue action raising irrigated farming's sector efficiency.
- Ponder long and hard about imposing adverse shocks on the irrigation sector, because of the devastation it can wreak on sustainable rural livelihoods.
- In the case where, in spite of all the counter-actions and counter-arguments, it is judged that allocation to irrigated agriculture should be reduced, regional government must develop its settlement, industrial and training infrastructure. Thereby new economic opportunities are opened up in villages, towns and cities to those who no longer engage in irrigated farming.

Such security exists where a country's agricultural sector has production capacity in a year of average harvests to meet in full the food needs of the resident population, or where the agricultural sector's exports in value terms are sufficient to permit the country to import all its residual food needs at the ruling level of world prices, or where the country's total exports are sufficient to permit the country to import all its residual food needs. Where national food security does exist, the sector efficiency argument for reallocation of water from agriculture to the household, industry and urban services sectors is reinforced. However, by extension of the sixteenth proposal above, national government should certainly review whether reallocation would be accompanied by a loss of export earnings, as well as gains, where the country either exports or could export agricultural produce and manufactures based on such products.

Alongside its long-term water resources strategy, the country should also review the prospects for its dollar exports of raw materials, manufactures and tourism services. The country should also review likely trends in the prices of imports, particularly food. If we have a case where reallocation will lead to losses of foreign exchange earnings, if non-food export prospects are not favourable and if there is a long-term upward trend forecast in the price on the international market of food imports such as wheat, maize and rice, then the good sense of water reallocation to urban uses may be open to strong challenge.

Index